Progress in Nonlinear Differential Equations and Their Applications
Volume 60

Brian H. Gilding
Robert Kersner

Travelling Waves in Nonlinear Diffusion-Convection Reaction

Springer Basel AG

Editors' addresses:

Brian H. Gilding
Department of Mathematics and Statistics
College of Science
Sultan Qaboos University
P.O. Box 36
Al-Khod 123
Sultanate of Oman

e-mail: gilding@squ.edu.om

Robert Kersner
Computer and Automation Research
Institute
Hungarian Academy of Sciences
P.O. Box 63
1518 Budapest
Hungary

e-mail: h1019ker@mail.iif.hu

2000 Mathematics Subject Classification 35K55, 35K57, 35K65, 35Qxx, 35R35, 92C15, 92C20, 92Dxx, 92E20

A CIP catalogue record for this book is available from the Library of Congress, Washington D.C., USA

Bibliographic information published by Die Deutsche Bibliothek
Die Deutsche Bibliothek lists this publication in the Deutsche Nationalbibliografie; detailed bibliographic data is available in the Internet at <http://dnb.ddb.de>.

ISBN 978-3-0348-9638-2 ISBN 978-3-0348-7964-4 (eBook)
DOI 10.1007/978-3-0348-7964-4

Originally published by Birkhäuser Verlag , Basel, Switzerland in 2004
Softcover reprint of the hardcover 1st edition 2004
Printed on acid-free paper produced of chlorine-free pulp. TCF ∞

ISBN 978-3-0348-9638-2

9 8 7 6 5 4 3 2 1 www.birkhasuer-science.com

Contents

Preface

This monograph has grown out of research we started in 1987, although the foundations were laid in the 1970's when both of us were working on our doctoral theses, trying to generalize the now classic paper of Oleinik, Kalashnikov and Chzhou on nonlinear degenerate diffusion. Brian worked under the guidance of Bert Peletier at the University of Sussex in Brighton, England, and, later at Delft University of Technology in the Netherlands on extending the earlier mathematics to include nonlinear convection; while Robert worked at Lomonosov State University in Moscow under the supervision of Anatolii Kalashnikov on generalizing the earlier mathematics to include nonlinear absorption. We first met at a conference held in Rome in 1985.

In 1987 we met again in Madrid at the invitation of Ildefonso Díaz, where we were both staying at 'La Residencia'. As providence would have it, the University 'Complutense' closed down during this visit in response to student demonstrations, and, we were very much left to our own devices. It was natural that we should gravitate to a research topic of common interest. This turned out to be the characterization of the phenomenon of finite speed of propagation for nonlinear reaction-convection-diffusion equations. Brian had just completed some work on this topic for nonlinear diffusion-convection, while Robert had earlier done the same for nonlinear diffusion-absorption. There was no question but that we bundle our efforts on the general situation.

What seemed to be a straightforward undertaking turned out to be a mammoth task, not so much because of the difficulty of the problem, but because of the wealth and diversity of the literature involved. With hindsight we were also too ambitious in that we wanted to show that our results encompassed all that had been done previously. Basically, our main result was that a nonlinear reaction-convection-diffusion equation displays finite speed of propagation if and only if it admits a travelling-wave solution with bounded support. In turn it admits such a travelling wave if and only if a singular integral equation involving the coefficients in the equation admits a solution satisfying a certain integrability criterion. The main results were announced in 1991. By this time though we had a manuscript which was obviously going to be difficult to get published because of its sheer length.

Our response was to submit it for publication in two pieces; one consisting of the analysis of the singular integral equation which appeared in 1993, and, the other dealing with the characterization of finite speed of propagation relying on the results in the first paper. The second component still turned out to be a fifty-three page article, which appeared in 1996.

Before the work on characterization appeared, however, it was clear that we had stumbled on something that we had missed initially. Namely, that we could dissect the questions of the admissibility of travelling-wave solutions with bounded support and the existence of a solution of the singular integral equation satisfying the integrability criterion. In essence, we could prove that a nonlinear reaction-convection-diffusion equation admits a travelling-wave solution if and only if the singular integral equation is solvable. Secondly, the travelling wave has bounded support if and only if the solution of the integral equation satisfies the integrability criterion. In writing this up, once again, we were overwhelmed by the literature on the topic, and, found our research paper growing out of reasonable proportion. At this juncture we decided to split the work yet again. The first publication would deal with the main results in the greatest possible mathematical generality. The companion would be more down to earth, review the literature on the topic, deal with applications, and, show how known results fall into the general theory.

The paper showing how the existence and uniqueness of travelling-wave solutions and wavefront solutions could be determined by studying the integral equation appeared in 1996. With regard to the companion, it transpired that we had still underestimated the task. The subject of travelling-wave solutions of diffusion-convection-reaction phenomena has seen an explosive growth in recent years. We could find some new result in, for instance, a physical or biological journal using an ad hoc technique or special features of phase-plane analysis; but no sooner had we shown how this fitted nicely into our theory, than we discovered yet another new result or application. What we had intended as an up-to-date research survey kept on growing, and in time attained a size which was too great even for consideration for publication in periodicals specializing in such reviews.

We are grateful to Birkhäuser Verlag for helping us out of the dilemma in which we found ourselves concerning the work. This monograph is the research survey in an alternative form. Keeping up with the literature and including everything on the topic finally turned out to be beyond our capabilities. Much recent literature has therefore unfortunately been omitted. This in no way reflects any judgement on its significance nor that we feel that we cannot capture it with our technique. We just had to stop somewhere. What we aim for in this monograph is to show how our integral equation technique can be successfully applied to obtain results which are otherwise inaccessible or improve on known results.

The book is organized as follows. The introduction reviews many well-known reaction-convection-diffusion equations to which our results apply, and, briefly describes the ideas and results. The second chapter describes the essence of our technique in full mathematical detail. It includes the theory of the integral equation which is needed later, and, presents the mathematical proofs of the key results describing a few of the technical hurdles. A number of straightforward corollaries of the theorems are also stated and proven in this chapter. The remainder of the monograph is then devoted to the application of the theorems to nonlinear reaction-convection-diffusion equations.

In Chapter 3, it is shown how known results on the equivalence of travelling-wave solutions of different reaction-convection-diffusion equations may be obtained and generalized. In Chapter 4, results are presented on how using the integral equation theory, the occurrence of a travelling-wave solution of a given equation may be invoked to deduce the existence of other such solutions of the same equation and of such solutions of other equations. Among other results, it is proven that if a reaction-convection-diffusion equation has a nontrivial travelling-wave solution with a certain wave speed, the same can be said for all greater wave speeds. Chapters 5 to 7 are then specifically concerned with the application of the integral equation theory to determine the existence of travelling-wave solutions of particular reaction-convection-diffusion equations. In Chapter 5, convection-diffusion equations are considered, and, in Chapter 6, reaction-diffusion equations. Chapter 7 is devoted to the power-law equation and a weak perturbation of the linear reaction-convection-diffusion equation involving logarithmic terms.

In Chapter 8 the existence of wavefront solutions of reaction-convection-diffusion equations is discussed and results on the set of wave speeds for which an equation may admit such a solution, analogous to the general results established beforehand, will be proven. In Chapter 9 such waves for convection-diffusion equations are characterized. Thereafter, in Chapter 10, similar waves for reaction-diffusion equations are studied.

For many, Chapter 10 may be the most interesting part of the book. Particular attention is paid to equations in which the reaction term has a fixed sign, to equations in which the reaction term has one sign change, and, to equations with smooth coefficients. Archetypes for the first category of equations are the Fisher, Newell–Whitehead, Zeldovich and KPP equations, while the Nagumo or bistable equation is the archetype for the second category of equations. Moreover, all of the afore-mentioned equations may be viewed as having smooth coefficients.

Chapter 11 concerns unbounded travelling-wave solutions. In Chapter 12 analyses of all global travelling-wave solutions and all unbounded travelling-wave solutions of the power-law equation are discussed.

The last chapter is devoted to explicit travelling-wave solutions which can be obtained from analysis of the integral equation.

The reader particularly interested in the use of the theory developed in this book could possibly find some of the mathematical proofs off-putting. For such a reader, we would recommend skipping the proofs, and, focussing on the contents of the theorems, the applications, and, the discussions surrounding both of these. A key feature of the theory is that once it has been developed, it is relatively straightforward to apply. The applications are taken from many areas.

Chapter 1

Introduction

Wave phenomena are observed in many natural reaction, convection and diffusion processes. This alone is motivation for studying their occurrence. Other reasons why the study of travelling-wave solutions has become such an essential part of the mathematical analysis of nonlinear diffusion-convection-reaction processes are that: the analysis of travelling waves provides a means of finding explicit solutions of the equation; in general travelling-wave solutions are easier to analyse and therewith discern properties to be expected of other solutions; such solutions can be used as tools in comparison principles and the like to determine the properties of general solutions; and, last but not least, in conformance with their natural occurrence in many mathematically modelled phenomena, they characterize the long-term behaviour in numerous situations.

This book concerns a technique which can be used to determine whether or not a nonlinear second-order scalar reaction-convection-diffusion equation admits a travelling-wave solution and to investigate the properties of such a travelling wave. The method involves the study of an integral equation, and is an alternative to phase-plane analysis. It will be applied to equations of the stated type, leading to extensions and improvements of existing results.

The class of partial differential equations that will be considered, contains prototypes which arise in many fields of application such as heat transfer, combustion, reaction chemistry, fluid dynamics, plasma physics, soil-moisture, foam drainage, crystal growth, biological population genetics, cellular ecology, neurology, and synergy. As a consequence the underlying interests and treatment are often different.

In contrast to much of the work on the topic in the past, in this monograph we shall consider equations which contain all of the following elements: nonlinear diffusion which may degenerate, nonlinear convection, and, nonlinear reaction which may be singular or discontinuous. Conventionally such equations are written in the form

$$u_t = (D(u)u_x)_x + B(u)u_x + F(u),$$

where subscripts denote partial differentiation, t denotes time, and x denotes a one-dimensional spatial variable. The first term on the right-hand side of this equation corresponds to a diffusive or dispersive process, with D the nonlinear diffusion coefficient. The second term on the right-hand side represents a convective or advective phenomenon, with B denoting a nonlinear velocity function. The last term corresponds to a reactive process, or sorption. If F is positive it represents a nonlinear source; if it is negative, a nonlinear sink. The unknown usually represents a nonnegative biological, physical or chemical variable such as density, saturation or concentration.

The following is a, by no means exhaustive, selection of equations that historically fall into the category above.

- The *Fisher equation* or *logistic equation*

$$u_t = u_{xx} + u(1-u)$$

which is the archetypical deterministic model for the spread of an advantageous gene in a population of diploid individuals living in a one-dimensional habitat.

- The *Newell–Whitehead equation* or *amplitude equation*

$$u_t = u_{xx} + u(1-u^2)$$

which arises after carrying out a suitable normalization in the study of thermal convection of a fluid heated from below. Considering the perturbation from a stationary state, the equation describes the evolution of the amplitude of the vertical velocity if this varies slowly.

- The *Zeldovich equation*

$$u_t = u_{xx} + u^2(1-u)$$

which arises in combustion theory. The unknown u represents temperature, while the last term on the right-hand side corresponds to the generation of heat by combustion.

- The *KPP equation*

$$u_t = u_{xx} + c(u),$$

with c differentiable for $0 \le u \le 1$, $c(0) = 0$, $c(u) > 0$ for $0 < u < 1$, $c(1) = 0$, and $c'(0) > c'(u)$ for $0 < u < 1$; of which the Fisher and Newell–Whitehead equations are special cases. This equation has the same origins as the Fisher equation and the Zeldovich equation respectively.

- The *Nagumo equation*or *bistable equation*

$$u_t = u_{xx} + u(1-u)(u-\alpha) \quad \text{with } 0 < \alpha < 1,$$

which has been obtained as one of a set of equations modelling the transmission of electrical pulses in a nerve axon.

- The *porous media equation* known in Soviet literature as the equation of Newtonian polytropic filtration

$$u_t = (u^m)_{xx} \quad \text{with } m > 0,$$

which reduces to the linear heat equation in the particular case $m = 1$. This equation has acquired its name because of its description of the flow of an adiabatic gas in a porous medium. The unknown u denotes the density of the gas and the constant m is related to its adiabatic constant. The equation also arises in other contexts. It can be used to describe nonlinear heat transfer, concentration-dependent diffusion, the motion of plasma particles in a magnetic field, and the evolution of biological populations. In the case $m = 2$ the equation can be found in boundary-layer theory, and as a dimensionless reformulation of the Boussinesq equation in hydrology. In the case $m = 7/2$ it arises in the study of solar prominences, while the case $m = 3/2$ occurs in the description of the imbibition of a wetting liquid in a capillary tube. The equation has even been proposed as a suitable model for the spread of intergalactic civilizations.

- The porous media equation with absorption $u_t = (u^m)_{xx} - u^p$ or with a source term $u_t = (u^m)_{xx} + u^p$ where $m > 0$ and $p > 0$.

- The *Burgers equation*
$$u_t + uu_x = u_{xx}.$$
This equation is famous as a model for the component of the velocity in one-dimensional turbulent flow.

- The porous media equation with convection $u_t = (u^m)_{xx} + (u^n)_x$ where m and n are positive constants. The *foam drainage equation*,
$$u_t = (u^{3/2})_{xx} + (u^2)_x,$$
is a particular example of this equation. Modelling the gravitational drainage of a foam comprising gas bubbles trapped in a liquid, the unknown in this example represents the liquid fraction. Another particular example is the equation $u_t = (u^4)_{xx} + (u^3)_x$. This example arises in the modelling of the motion of a thin sheet of viscous liquid over an inclined plate. The unknown in this model represents the film thickness. In both examples t denotes time and x a distance which decreases in the direction of gravitational pull.

- The *Richards equation* also referred to as the *nonlinear Fokker–Planck equation*
$$u_t = (D(u)u_x + b(u))_x.$$
Under appropriate conditions on the functions D and b, this equation models the one-dimensional transport of water in an unsaturated homogeneous soil. In this context u denotes soil-moisture content. The mechanism behind the second-order term on the right-hand side of the equation is capillary suction, while the first-order term is related to the influence of gravity and is proportional to the hydraulic conductivity of the soil.

- The so-called quenching problem

$$u_t = (u^m)_{xx} - \begin{cases} u^{-p} & \text{for } u > 0 \\ 0 & \text{for } u = 0 \end{cases}$$

with $m > 0$ and $p > 0$.

- Combustion models with ignition thresholds

$$u_t = u_{xx} + \begin{cases} 0 & \text{for } 0 \le u < \delta \\ c(u) & \text{for } u \ge \delta \end{cases}$$

where $c(u) > 0$ for $u > \delta$ and $c(\delta) \ge 0$. Such equations describe the deflagration of a flame with one reactant in a single-step chemical reaction. The unknown u denotes the normalized temperature, and $c(u)$ a normalized reaction term with ignition temperature δ.

- The porous media equation with convection and reaction

$$u_t = (u^m)_{xx} + (b_0 u^n)_x + c_0 u^p - c_1 u^q,$$

where $m > 0$, $n > 0$, $p > 0$, $q > 0$, b_0, $c_0 \ge 0$ and $c_1 \ge 0$ are constants.

Following the mathematical tradition of writing the porous media equation and its generalizations in the style above, we shall consider all of the above equations, and many more besides, as specific examples of equations of the general class

$$u_t = (a(u))_{xx} + (b(u))_x + c(u). \tag{1.1}$$

In this formulation a' plays the role of the diffusion coefficient. The coefficient b may be viewed as a nonlinear convective flux function, with b' the corresponding nonlinear velocity.

In correspondence with the predominant modelling origins of equation (1.1), only nonnegative solutions of the equation will be considered. However, these solutions may be bounded or unbounded. In contrast to much of the previous work, we shall consider hypotheses on the coefficients a, b and c of such a general nature that they may be regarded as lying on the border of the current requirements for a mathematical theory of partial differential equations of the form (1.1). We shall allow a' and b' to vanish or become singular at certain points, and the reaction term c to be discontinuous. The precise details will be presented in the next chapter.

Suppose that equation (1.1) admits a travelling-wave solution of the form

$$u(x,t) = f(\xi) \quad \text{with} \quad \xi = x - \sigma t \tag{1.2}$$

where σ is a constant which constitutes the *wave speed*. Then formally substituting (1.2) into (1.1) yields the ordinary differential equation

$$(a(f))'' + (b(f))' + c(f) + \sigma f' = 0 \tag{1.3}$$

where a prime denotes differentiation with respect to ξ.

A *wavefront* solution of equation (1.1) is a particular type of travelling-wave solution that is defined for all ξ, monotonic, and such that f connects two equilibrium states of the equation. Typically,

$$f(\xi) \to 0 \quad \text{as } \xi \to \infty \tag{1.4}$$

and

$$f(\xi) \to \ell \quad \text{as } \xi \to -\infty, \tag{1.5}$$

where 0 and $\ell \neq 0$ are equilibrium points of the reaction term, i.e. $c(0) = c(\ell) = 0$.

In general, for the class of partial differential equations considered, the ordinary differential equation (1.3) does not have a classical solution. For instance in the case of the porous media equation, $u_t = (u^m)_{xx}$, equation (1.3) reads $(f^m)'' + \sigma f' = 0$. For $m > 1$ this admits the solution

$$f(\xi) = \begin{cases} \left(-\frac{m-1}{m}\sigma\xi\right)^{1/(m-1)} & \text{for } \xi < 0 \\ 0 & \text{for } \xi \geq 0 \end{cases} \tag{1.6}$$

for every $\sigma > 0$. This solution is physically relevant, since f and $(f^m)'$, which, in the process of gas flow in a porous medium modelled, correspond to the density and flux respectively, are continuous. However, when $m \geq 2$, the given function f is not continuously differentiable at $\xi = 0$ and thus does not constitute a classical solution.

Note too that if f is a travelling-wave solution satisfying (1.4), this condition does not automatically mean that f is positive and approaches zero in the limit. In many cases the typical behaviour is $f(\xi) = 0$ for all $\xi \geq \xi^*$ for some argument ξ^*. This is illustrated by the explicit solution (1.6) of the porous media equation with $m > 1$. A similar remark may be made with regard to the condition (1.5).

Setting $g(\xi) := -(a(f))'(\xi)$ the ordinary differential equation (1.3) can be reformulated as the first-order system

$$\begin{cases} f' &= -g/a'(f) \\ g' &= -g\{\sigma + b'(f)\}/a'(f) + c(f). \end{cases} \tag{1.7}$$

In the phase-plane the trajectories of this system are given by

$$\frac{dg}{df} = \sigma + b'(f) - \frac{c(f)a'(f)}{g}. \tag{1.8}$$

Subsequently, if (1.3) admits a solution for which (1.4) holds and

$$(a(f))'(\xi) \to 0 \quad \text{as } \xi \to \infty,$$

this solution is necessarily represented by a trajectory which approaches the point $(f, g) = (0, 0)$ in the phase-plane. Hence, by setting

$$\theta(f(\xi)) = g(\xi) \quad \text{and} \quad s = f(\xi)$$

and integrating (1.8) through $(0,0)$, such a trajectory may be characterized as a solution of the integral equation

$$\theta(s) = \sigma s + b(s) - \int_0^s \frac{c(r)a'(r)}{\theta(r)}\,dr \tag{1.9}$$

in which θ is the unknown. Conversely, retracing the above argument, a solution f of (1.3) can be constructed from a solution θ of the integral equation by solving

$$\frac{a'(f)}{\theta(f)} f' = -1. \tag{1.10}$$

The purpose of this book is to show that by studying the integral equation (1.9) as opposed to the traditional phase-plane, one may obtain new insights and an extended theory for travelling-wave solutions of the nonlinear reaction-convection-diffusion equation (1.1).

It is seldom that practitioners (even mathematicians) are able to give a precise global phase-plane analysis of the system of nonlinear ordinary differential equations (1.7). One of the advantages of using the integral equation approach lies in the fact that in practice one has only to check a few particulars of the equation. Moreover, in concrete situations, this is easily done. To analyse the more detailed behaviour of travelling waves, it remains to understand the geometric properties of solutions θ of the integral equation. These may differ from application to application.

Taking into account a condition on θ which naturally allows one to find f from (1.10), and supposing that $c(0) = c(\ell) = 0$ with $\ell > 0$, one of the many results which illustrates the power of the approach is the following.

> Equation (1.1) admits a wavefront solution with speed σ satisfying (1.4) and (1.5) if and only if the integral equation (1.9) has a solution θ on $[0,\ell]$ such that $\theta(\ell) = 0$.

Similar results cover other kinds of travelling waves.

By studying the integral equation (1.9) and applying results like the one above, previous results on the occurrence of travelling-wave solutions of equations of the class (1.1) will be unified and generalized in this book. It will be shown how the integral equation may be applied in a number of different concrete situations.

By this approach, questions involving spectra of wave speeds, and, minimal wave speeds can be handled. We shall see its close link with the so-called variational method. Easy-to-check necessary and sufficient conditions for the existence of travelling-wave solutions may given in terms of the coefficients in the reaction-diffusion-convection equation. Some transformations which may otherwise appear somewhat mysterious can be explained in a natural way. Last but not least, the method permits one to find explicit solutions in a rather simple fashion.

Chapter 2

General theory

2.1 Basic hypotheses

Motivated by the modelling origins of equation (1.1), where the unknown is generally nonnegative, we shall henceforth assume that the coefficients of the equation are defined in some closed interval I with minimum 0. However, because we would like to deal with both bounded and unbounded solutions in a single framework, let us provisionally denote the supremum of this interval by ℓ which may be finite or infinite. Thus,

$$I = [0, \ell) \quad \text{with } \ell = \infty, \qquad \text{or,} \qquad I = [0, \ell] \quad \text{with } 0 < \ell < \infty.$$

The assumptions on equation (1.1) are the following.

Hypothesis 2.1. The coefficients a, b and c are defined on I and real. Furthermore:

(i) The function a is continuous in I, continuously differentiable in $(0, \ell)$, $a'(u) > 0$ for $0 < u < \ell$, and $a(0) = 0$.

(ii) The function b is continuous in I, differentiable in $(0, \ell)$, and $b(0) = 0$.

(iii) The function c is such that the one-sided limits $c(u-)$ and $c(u+)$ exist for every $0 < u < \ell$, with $c(u-) < 0$ if $c(u) < 0$, and, $c(u+) > 0$ if $c(u) > 0$;

$$\int_0^s |c(r)|\, a'(r)\, dr < \infty \qquad \text{for all } 0 < s \le \ell, \quad s < \infty;$$

$c(0) = 0$; and, $c(\ell) = 0$ if $\ell < \infty$.

This hypothesis is met by all the model equations reviewed in the previous chapter, with the proviso that $p < m$ in the so-called quenching problem in order to meet condition (iii). Moreover, it is of such a general nature that it may be regarded as lying on the border of the current requirements for a mathematical theory for partial differential equations of the form (1.1).

The example (1.6) discussed in the introduction illustrates the necessity of the consideration of some abstraction of the notion of a solution. For continuous c in I it would suffice to consider weak solutions of equation (1.3). However, one would like to be able to consider singular reaction terms such as that in the so-called quenching problem. We therefore introduce a definition of a solution of (1.3) that permits consideration of travelling waves that do not even need to be such that $c(f)$ is everywhere locally integrable. To this end, in the event that $\ell = \infty$ we let

$$c_\varepsilon(s) = \begin{cases} 0 & \text{for } s \leq \varepsilon \\ c(s) & \text{for } s > \varepsilon \end{cases}$$

for any $\varepsilon > 0$, and if $\ell < \infty$ is bounded

$$c_\varepsilon(s) = \begin{cases} 0 & \text{for } s \leq \varepsilon \\ c(s) & \text{for } \varepsilon < s < \ell - \varepsilon \\ 0 & \text{for } s \geq \ell - \varepsilon \end{cases}$$

for any $\ell/2 > \varepsilon > 0$.

Definition 2.2. A function f defined on an open real interval Ω with values in I is said to be a travelling-wave solution of equation (1.1) with speed σ if $f \in C(\Omega)$, $(a(f))' \in L^1_{\text{loc}}(\Omega)$, $c_\varepsilon(f) \in L^1_{\text{loc}}(\Omega)$ for all sufficiently small $\varepsilon > 0$, and

$$\lim_{\varepsilon \downarrow 0} \int_\Omega [\{(a(f))' + b(f) + \sigma f\}\phi' - c_\varepsilon(f)\phi]\, d\xi = 0 \tag{2.1}$$

for any function $\phi \in C_0^\infty(\Omega)$.

Any classical or weak solution of the ordinary differential equation (1.3) in an interval Ω is automatically a solution in the above sense.

Definition 2.3. If, in the previous definition, $\Omega = (-\infty, \infty)$, the function f is said to be a *global* travelling-wave solution of equation (1.1).

Definition 2.4. Two travelling-wave solutions will be said to be *indistinct* if one is a translation of the other or one is the restriction of the other to a smaller domain. Otherwise they are *distinct*.

Definition 2.5. A bounded global travelling-wave solution that is monotonic, but not constant, and such that

$$f(\xi) \to \ell^- \quad \text{as } \xi \to -\infty \tag{2.2}$$

and

$$f(\xi) \to \ell^+ \quad \text{as } \xi \to \infty \tag{2.3}$$

for some $\ell^-, \ell^+ \in I$ with $c(\ell^-) = c(\ell^+) = 0$ is said to be a *wavefront*solution from ℓ^- to ℓ^+.

Definition 2.6. A travelling-wave solution that is defined, monotonic, but not constant in (ω, ∞) for some real ω, and such that (2.3) holds for some $\ell^+ \in I$ with $c(\ell^+) = 0$ is said to be a *semi-wavefront solution to* ℓ^+. Similarly, a travelling-wave solution that is defined, monotonic, but not constant in $(-\infty, \omega)$ for some real ω, and such that (2.2) holds for some value ℓ^- with $c(\ell^-) = 0$ is said to be a *semi-wavefront solution from* ℓ^-.

The function $f(\xi) = \{(1-m)\sigma\xi/m\}^{-1/(1-m)}$ for any $\sigma > 0$ gives an example of a semi-wavefront solution for the porous media equation $u_t = (u^m)_{xx}$ with $m < 1$. For $\xi > 0$ it can be verified to satisfy $(f^m)'' + \sigma f' = 0$ classically, be monotonic decreasing, and, such that $f(\xi) \to 0$ as $\xi \to \infty$. However, it is not extendible beyond the interval $(0, \infty)$.

Definition 2.7. A semi-wavefront solution that is not extendible to a global travelling-wave solution is said to be a *strict semi-wavefront* solution.

The interest in this monograph will be in semi-wavefront, wavefront and unbounded monotonic travelling-wave solutions of equations of the class (1.1). Under favourable conditions on the coefficients in the equation it can be shown that any global travelling-wave solution of such an equation satisfying (2.2) and (2.3) and with values in the range $\min\{\ell^-, \ell^+\} \le f \le \max\{\ell^-, \ell^+\}$ is necessarily monotonic [102,103]. Moreover, it can be established that, apart from the constant solutions which are easy to identify, the only bounded monotonic global travelling-wave solutions of an equation of the type (1.1) are its wavefront solutions. We shall return to both these points in Section 2.6.

To supplement the above remarks, we note that by a simple change of variables any wavefront solution of an equation of the type (1.1) can be transformed into a wavefront solution from ℓ to 0 for an equivalent equation of the same class. Similarly, any semi-wavefront solution can be transformed into a semi-wavefront solution decreasing to 0. We shall therefore, without loss of generality, focus our attention on wavefront solutions from $\ell < \infty$ to 0 and semi-wavefront solutions to 0.

As stated in the introduction the purpose of this book is to show that the study of monotonic travelling-wave solutions of equation (1.1) is equivalent to the study of the integral equation (1.9) in which θ is the unknown. This equation may be classified as a singular nonlinear Volterra integral equation of the second kind [136,204]. The theory for it has been developed in [123].

To fully describe the correspondence between travelling-wave solutions of (1.1) and solutions of (1.9) however, we need an additional constraint. Returning to the derivation of equation (1.9) in the introduction, we see that a travelling-wave solution f and a solution θ of the integral equation, are to be related formally by (1.10). This infers that in actuality

$$\int_{\nu}^{f(\xi)} \frac{a'(r)}{\theta(r)}\, dr = \xi_0 - \xi$$

for some ν in the domain of definition of θ and some number ξ_0. To justify this construction, the above integral must be well-defined. To be specific, for a travelling-wave solution of equation (1.1) with values $f(\xi)$ in $(0, \delta)$, this means that necessarily θ is defined in the domain $(0, \delta)$, and,

$$\int_{s_0}^{s_1} \frac{a'(r)}{\theta(r)}\, dr < \infty \qquad \text{for all } 0 < s_0 < s_1 < \delta. \tag{2.4}$$

Henceforth, we shall refer to this constraint as follows.

Definition 2.8. A solution θ of equation (1.9) in a right neighbourhood of zero $[0,\delta)$, or $[0,\delta]$ with $\delta < \infty$, is said to satisfy the *integrability condition* in this neighbourhood if (2.4) holds.

With this proviso, monotonic travelling-wave solutions of equation (1.1) are characterized by solutions of the integral equation (1.9). The following can be proven.

Theorem 2.9. *Equation* (1.1) *has a semi-wavefront solution with speed* σ *decreasing to* 0 *if and only if* (1.9) *has a solution satisfying the integrability condition on an interval* $[0,\delta)$ *for some* $0 < \delta \leq \ell$.

Theorem 2.10. *Suppose that* $\ell < \infty$. *Then equation* (1.1) *has a wavefront solution from* ℓ *to* 0 *with speed* σ *if and only if* (1.9) *has a solution* θ *satisfying the integrability condition on* $[0,\ell]$ *such that* $\theta(\ell) = 0$.

Theorem 2.11. *Suppose that* $\ell = \infty$. *Then equation* (1.1) *has an unbounded monotonic travelling-wave solution with speed* σ *decreasing to* 0 *if and only if* (1.9) *has a solution* θ *satisfying the integrability condition on* $[0,\infty)$.

2.2 Integral equation theory

The difficulty with any analysis of the integral equation (1.9) lies in its singular kernel. Consider the more general equation

$$\theta(s) = \mathsf{F}(s) - \int_0^s \frac{g(r)}{\theta(r)}\,dr, \tag{2.5}$$

where F is continuous in I with $\mathsf{F}(0) \geq 0$, and, g is integrable in every bounded subset of I. Let

$$\mathsf{G}(r,\theta) = \begin{cases} g(r)/\theta & \text{if } \theta > 0 \\ -\infty & \text{if } g(r) < 0 \text{ and } \theta = 0 \\ 0 & \text{if } g(r) = 0 \text{ and } \theta = 0 \\ \infty & \text{if } g(r) > 0 \text{ and } \theta = 0. \end{cases}$$

Definition 2.12. A function θ is said to be a solution of equation (2.5) in a right neighbourhood of zero $[0,\delta) \subseteq I$ if it is defined, real, nonnegative and continuous on $[0,\delta)$, $\mathsf{G}(r,\theta(r))$ is integrable on every compact subset of $(0,\delta)$,

$$\int_0^s \mathsf{G}(r,\theta(r))\,dr := \lim_{\varepsilon\downarrow 0}\int_\varepsilon^s \mathsf{G}(r,\theta(r))\,dr \quad \text{exists}$$

and satisfies

$$\theta(s) = \mathsf{F}(s) - \int_0^s \mathsf{G}(r,\theta(r))\,dr \tag{2.6}$$

for all $s \in (0, \delta)$. A function θ is said to be a *solution of* (2.5) *in a compact right neighbourhood of zero* $[0, \delta]$ if in addition θ is continuous on $[0, \delta]$ and one may pass to the limit $s \uparrow \delta$ in (2.6).

Results from the theory of equation (2.5) which were proven in [123] and will be applied to equation (1.9) in this monograph include the following. In a number of these results an important role is played by the variable

$$Q(s) := \left| 2 \int_0^s g(r)\, dr \right|^{1/2}.$$

Lemma 2.13 (The nonsingular case). *If* $\mathsf{F}(0) > 0$ *equation* (2.5) *has a unique positive solution* θ *in a right neighbourhood of zero* $[0, \delta)$ *such that* $\delta = \ell$ *or* $\theta(s) \to 0$ *as* $s \uparrow \delta$.

Lemma 2.14 (Uniqueness). *Equation* (2.5) *has no solution, a unique solution, or, an uncountable number of solutions which are majorized by a unique maximal solution. Moreover:*

(i) *If* $g(s) \leq 0$ *for all* $s \in (0, \delta)$ *some* $0 < \delta \leq \ell$, *then either* (2.5) *has no solution in* $[0, \delta)$ *or it has a unique solution in* $[0, \delta)$.

(ii) *If* $g(s) \geq 0$ *and* $Q(s) > 0$ *for all* $s \in (0, \delta)$ *some* $0 < \delta \leq \ell$ *and given any* $s \in (0, \delta)$ *there is an* $s^* \in (s, \delta)$ *and a* $0 < k < 2$ *such that* $\mathsf{F}(r) - \mathsf{F}(s) \leq k\{2 \int_s^r g(w)\, dw\}^{1/2}$ *for all* $r \in (s, s^*)$, *then either* (2.5) *has no solution in* $[0, \delta)$ *or it has an uncountable number of solutions in* $[0, \delta)$.

Lemma 2.15 (Existence with a nonnegative integrand). *Suppose that* $g(s) \geq 0$ *and* $Q(s) > 0$ *for all* $s \in (0, \delta)$ *some* $0 < \delta \leq \ell$. *Define*

$$L(s) := 1/\left|\ln Q(s)\right| \quad \text{and} \quad J(s) := 1/\left|\ln L(s)\right|. \tag{2.7}$$

(i) *If* $\mathsf{F}(s) \geq (2Q - QL^2\{1 + J^2\}/4)(s)$ *for all* $s \in (0, \delta)$, *then* (2.5) *admits a maximal solution* θ *in a right neighbourhood of zero. Moreover,* $\mathsf{F}(s) - (Q + QL\{1 + J\}^2/2)(s) \leq \theta(s) \leq \mathsf{F}(s)$ *for all* $s \in (0, \delta^*)$ *for some* $0 < \delta^* \leq \delta$.

(ii) *If* $\mathsf{F}(s) \leq (2Q - kQL^2)(s)$ *for all* $s \in (0, \delta)$ *for some* $k > 1/4$, *then* (2.5) *has no solution in* $[0, \delta)$.

Lemma 2.16 (Existence with a nonpositive integrand). *Suppose that* $g(s) \leq 0$ *and* $Q(s) > 0$ *for all* $s \in (0, \delta)$ *some* $0 < \delta \leq \ell$.

(i) *If for any* $s \in (0, \delta)$ *there is an* $s^* \in (0, s)$ *and a* $0 < k < 2$ *such that* $\mathsf{F}(r) - \mathsf{F}(s) \leq k\{2 \int_r^s |g(w)|\, dw\}^{1/2}$ *for all* $r \in (s^*, s)$, *then* (2.5) *has a unique solution* θ *in* $[0, \delta)$. *Moreover, if there exists a point* $s \in (0, \delta)$ *for which* $\theta(s) = 0$, *then* $\theta \equiv Q \equiv 0$ *in* $[0, s]$.

(ii) *If* $\alpha Q(s) \leq \mathsf{F}(s) \leq \beta Q(s)$ *for all* $s \in (0, \delta)$ *for some constants* $-\infty < \alpha \leq \beta \leq \infty$ *with* $\alpha \geq -2/(\sqrt{\beta^2 + 4} + \beta)$, *then* (2.5) *has a unique solution* θ *in* $[0, \delta)$. *Moreover, for all* $s \in [0, \delta)$,

$$\mathsf{F}(s) + \{2/(\sqrt{\beta^2 + 4} + \beta)\}Q(s) \leq \theta(s) \leq \mathsf{F}(s) + \{2/(\sqrt{\alpha^2 + 4} + \alpha)\}Q(s).$$

Lemma 2.17 (Extendibility). *Any solution θ of equation (2.5) in a bounded interval $[0,\delta) \subseteq I$ is a solution of (2.5) in $[0,\delta]$. Moreover, $\delta = \ell$, $\theta(s) \to 0$ as $s \uparrow \delta$, or, θ is extendible onto an interval $[0,\delta^*)$ with $\delta < \delta^* \leq \ell$.*

Lemma 2.18 (Comparison principles). *Consider equation (2.5) with two different sets of parameters and coefficients F_i and g_i on an interval I for $i = 1, 2$.*

(i) *Suppose that $\mathsf{F}_2(0) \geq \mathsf{F}_1(0)$, $s \mapsto \mathsf{F}_2(s) - \mathsf{F}_1(s)$ is nondecreasing on $[0,\delta)$, and, $g_2 \leq g_1$ almost everywhere in $(0,\delta)$, for some $0 < \delta \leq \ell$. Then, if (2.5) with $i = 1$ has a solution θ_1 in $[0,\delta)$ such that $g_2/\theta_1 \in L^1_{\mathrm{loc}}(0,\delta)$, the equation with $i = 2$ has a solution θ_2 in $[0,\delta)$ such that $\theta_2(s) \geq \theta_1(s)$ for all $s \in [0,\delta)$. Moreover, if $\mathsf{F}_2(s) - \mathsf{F}_1(s) > \mathsf{F}_2(r) - \mathsf{F}_1(r)$ for all $0 \leq r < s$, then $\theta_2(s) = \theta_1(s)$ if and only if $\theta_2(s) = 0$.*

(ii) *Suppose that $\mathsf{F}_2 \geq \mathsf{F}_1$ on $[0,\delta)$, and, $\max\{0, g_2\} \leq g_1$ almost everywhere in $(0,\delta)$ for some $0 < \delta \leq \ell$. Then, if (2.5) with $i = 1$ has a solution θ_1 in $[0,\delta)$, the equation with $i = 2$ has a solution θ_2 in $[0,\delta)$ such that $\theta_2(s) \geq \theta_1(s) + \mathsf{F}_2(s) - \mathsf{F}_1(s)$ for all $s \in [0,\delta)$.*

Lemma 2.19 (An explicit case). *Suppose that $g(s) \geq 0$, $Q(s) > 0$ and $\mathsf{F}(s) = kQ(s)$ for all $s \in (0,\delta)$ for some k and $0 < \delta < \ell$. Define $L(s)$ and $J(s)$ by (2.7).*

(i) *If $k = 2$, then (2.5) admits the maximal solution $\theta(s) = Q(s)$ in $[0,\delta)$, for every real number γ a unique solution θ_γ in a right neighbourhood of zero such that $\theta_\gamma(s) = (Q - QL + QL^2J^{-1} + \gamma QL^2)(s) + \mathcal{O}((QL^3J^{-2})(s))$ as $s \downarrow 0$, and no other solutions. Moreover, if $[0,\delta_\gamma)$ denotes the maximal subinterval of $[0,\delta)$ in which θ_γ exists, then $\delta_\gamma = \sup\{s \in [0,\delta) : Q(r) \leq \exp(\gamma - 1)$ for all $r \in [0,s)\}$.*

(ii) *If $k > 2$, then setting $\beta_1 := (k - \sqrt{k^2-4})/2$ and $\beta_2 := (k + \sqrt{k^2-4})/2$, equation (2.5) admits the maximal solution $\theta(s) = \beta_2 Q(s)$ in $[0,\delta)$, for every real number γ a unique solution θ_γ in a right neighbourhood of zero such that $\theta_\gamma(s) = (\beta_1 Q + \gamma Q^{\beta_2/\beta_1})(s) + \mathcal{O}(Q^{(2\beta_2-\beta_1)/\beta_1}(s))$ as $s \downarrow 0$, and no other solutions. Moreover, if $[0,\delta_\gamma)$ denotes the maximal subinterval of $[0,\delta)$ in which θ_γ exists, then for $\gamma > 0$ one has $\delta_\gamma = \delta$, and $\theta_\gamma(s) \sim \beta_2 Q(s)$ as $s \uparrow \delta$ if $Q(s) \to \infty$ as $s \uparrow \delta$; one has $\delta_0 = \delta$ and $\theta_0(s) = \beta_1 Q(s)$ for all $s \in [0,\delta)$; and finally for $\gamma < 0$ one has $\delta_\gamma = \sup\{s \in [0,\delta) : Q^{\beta_2-\beta_1}(r) \leq \beta_1^{\beta_1}\beta_2^{-\beta_2}(\beta_2-\beta_1)^{\beta_2}|\gamma|^{-\beta_1}$ for all $r \in [0,s)\}$.*

In both cases $\theta(s) > \theta_{\gamma^*}(s) > \theta_\gamma(s)$ for all $s \in (0,\delta_\gamma)$ and $-\infty < \gamma < \gamma^* < \infty$.

2.3 Proof of equivalence

Let us begin with some preliminary information on travelling-wave solutions of (1.1).

Lemma 2.20. *Suppose that (1.1) has a travelling-wave solution f in an open interval Ω.*

(i) *If* $f \equiv f^*$ *for some* $f^* \in I$, *then* $c(f^*) = 0$.

(ii) *If* $0 < f(\xi) < \ell$ *for all* $\xi \in \Omega$, *then* $a(f) \in C^1(\Omega)$, $(a(f))' + b(f) + \sigma f$ *is absolutely continuous in* Ω, *and*

$$((a(f))' + b(f) + \sigma f)' + c(f) = 0 \tag{2.8}$$

almost everywhere in Ω.

(iii) *If* $\ell > f(\xi_1) > f(\xi) > 0$ *for all* $\xi \in (\xi_1, \xi_2) \subseteq \Omega$ *and some* $\xi_1 \in \Omega$, *then* $(a(f))'(\xi_1) < 0$ *or* $c(f(\xi_1)) \geq 0$.

(iv) *If* $\ell > f(\xi) > f(\xi_1) > 0$ *for all* $\xi \in (\xi_0, \xi_1) \subseteq \Omega$ *and some* $\xi_1 \in \Omega$, *then* $(a(f))'(\xi_1) < 0$ *or* $c(f(\xi_1)) \leq 0$.

Proof. Part (i) may be verified formally by substitution in (1.3), and follows rigorously from Definition 2.2 of a travelling-wave solution of equation (1.1). With regard to part (ii), when $0 < f < \ell$, Definition 2.2 is equivalent to the definition of a solution of (1.3) in the sense of distributions. The assertion may subsequently be obtained from a standard regularity argument. As regards parts (iii) and (iv) in the classical setting, if the ordinary differential equation (1.3) has a local maximum in a point ξ_1, then by substitution in (1.3) necessarily $c(f(\xi_1)) \geq 0$, whereas if this equation has a local minimum in ξ_1, then $c(f(\xi_1)) \leq 0$. Parts (iii) and (iv) of the lemma are the extension of this argument to the present situation with weak continuity assumptions on the coefficients in equation (1.1). We refer to [124] for details. □

With the above information we can prove the 'only if' statements of Theorems 2.9, 2.10 and 2.11.

Suppose that equation (1.1) has a monotonic travelling-wave solution in an interval $\Omega = (\omega, \infty)$ for which $f(\xi) \to 0$ as $\xi \to \infty$ and $f(\xi) \to \delta$ as $\xi \to \omega$ with $0 < \delta \leq \ell$ and $-\infty \leq \omega < \infty$. Set

$$\Xi_0 := \sup\{\xi \in (\omega, \infty) : f(\xi) > 0\}$$

and

$$\Xi_1 := \inf\{\xi \in (\omega, \infty) : f(\xi) < \delta\}.$$

By Lemma 2.20(ii), $a(f) \in C^1(\Xi_1, \Xi_0)$, $(a(f))' + b(f) + \sigma f$ is absolutely continuous on (Ξ_1, Ξ_0) and (2.8) holds almost everywhere in (Ξ_1, Ξ_0). Subsequently, one can define a continuous nonnegative function θ on $(0, \delta)$ via

$$\theta(f(\xi)) = -(a(f))'(\xi) \quad \text{for } \xi \in (\Xi_1, \Xi_0). \tag{2.9}$$

We shall show that θ is the solution of the integral equation (1.9) satisfying the integrability condition (2.4) which we seek.

Let

$$S := \{\xi \in (\Xi_1, \Xi_0) : (a(f))'(\xi) = 0\},$$

and note that by Lemma 2.20, $c(f(\xi)) = 0$ for all $\xi \in S$. Consequently,

$$\int_{\xi_1}^{\xi_2} |c(f(\xi))|\, d\xi = \int_{(\xi_1,\xi_2)\setminus S} |c(f(\xi))|\, d\xi = \lim_{\varepsilon\downarrow 0} \int_{\xi_1}^{\xi_2} \frac{-|c(f(\xi))|\,(a(f))'(\xi)}{\theta(f(\xi))+\varepsilon}\, d\xi$$

for any $\Xi_1 < \xi_1 < \xi_2 < \Xi_0$. Hence, with the convention that any fraction of the form ca'/θ takes the value 0 if both ca' and θ vanish,

$$\int_{\xi_1}^{\xi_2} |c(f(\xi))|\, d\xi = \lim_{\varepsilon\downarrow 0} \int_{f(\xi_2)}^{f(\xi_1)} \frac{|c(r)|\, a'(r)}{\theta(r)+\varepsilon}\, dr = \int_{f(\xi_2)}^{f(\xi_1)} \left|\frac{c(r)a'(r)}{\theta(r)}\right| dr$$

for any $\Xi_1 < \xi_1 < \xi_2 < \Xi_0$. It follows that with the above convention, ca'/θ is integrable on every compact subset of $(0, \delta)$. Furthermore, repeating the above argument without the absolute value signs

$$\int_{f(\xi_2)}^{f(\xi_1)} \frac{c(r)a'(r)}{\theta(r)}\, dr = \int_{\xi_1}^{\xi_2} c(f(\xi))\, d\xi \quad \text{for all } \Xi_1 < \xi_1 < \xi_2 < \Xi_0. \tag{2.10}$$

Alternatively repeating the above argument with c replaced by 1 we deduce that a'/θ is integrable on every compact subset of $(0, \delta)$, and

$$\int_{f(\xi_2)}^{f(\xi_1)} \frac{a'(r)}{\theta(r)}\, dr = \xi_2 - \xi_1 - \int_{(\xi_1,\xi_2)\cap S} 1\, d\xi \tag{2.11}$$

for all $\Xi_1 < \xi_1 < \xi_2 < \Xi_0$.

Integrating (2.8) and combining with (2.9) and (2.10), θ satisfies the equation

$$\theta(s_1) - \sigma s_1 - b(s_1) = \theta(s_0) - \sigma s_0 - b(s_0) - \int_{s_0}^{s_1} \frac{c(r)a'(r)}{\theta(r)}\, dr$$

for any $0 < s_0 < s_1 < \delta$. By Lemma 2.17, this implies

$$\mu := \lim_{s\downarrow 0} \theta(s) \geq 0 \quad \text{exists,} \tag{2.12}$$

$$\int_0^s \frac{c(r)a'(r)}{\theta(r)}\, dr := \lim_{\varepsilon\downarrow 0} \int_\varepsilon^s \frac{c(r)a'(r)}{\theta(r)}\, dr \quad \text{exists,}$$

and

$$\theta(s) = \mu + \sigma s + b(s) - \int_0^s \frac{c(r)a'(r)}{\theta(r)}\, dr$$

for any $0 < s < \delta$. Whence, if $\mu = 0$, θ satisfies the integral equation (1.9) and the integrability condition (2.4) on $[0, \delta)$.

It remains to show that $\mu = 0$. However, amalgamating (2.9) and (2.12) implies $(a(f))'(\xi) \to -\mu$ as $\xi \uparrow \Xi_0$. Hence, if $\Xi_0 = \infty$, this can plainly only be the case if $\mu = 0$. Whereas if $\Xi_0 < \infty$, since $(a(f))' \equiv f \equiv 0$ on (Ξ_0, ∞), from the integral identity (2.1) in the definition of a travelling-wave solution of (1.1) on (Ξ_1, ∞) it can be verified that necessarily $\mu = 0$.

This supplies the 'only if' conclusions of Theorems 2.9 and 2.11. The 'only if' conclusions of Theorem 2.10 may be proven by extending the above considerations regarding the limit (2.12) to the corresponding limit as $s \uparrow \ell$.

To verify the 'if' conclusions of Theorems 2.9 to 2.11, suppose now that the integral equation (1.9) has a solution θ satisfying the integrability condition on an interval $[0, \delta)$ with $0 < \delta \leq \ell$. Suppose furthermore that if $\delta = \ell < \infty$, then θ solves (1.9) on $[0, \ell]$, and $\theta(\ell) = 0$. Let $\nu \in (0, \delta)$ and set

$$\Xi_0 := \int_0^\nu \frac{a'(r)}{\theta(r)}\, dr \quad \text{and} \quad \Xi_1 := -\int_\nu^\delta \frac{a'(r)}{\theta(r)}\, dr. \tag{2.13}$$

Next, let Ω denote the interval $(-\infty, \infty)$ if $\delta = \ell < \infty$ and the interval (Ξ_1, ∞) otherwise. Finally, define the function f on Ω by

$$-\int_\nu^{f(\xi)} \frac{a'(r)}{\theta(r)}\, dr = \max\{\xi, \Xi_1\} \quad \text{for } \xi \leq 0 \tag{2.14}$$

and

$$\int_{f(\xi)}^\nu \frac{a'(r)}{\theta(r)}\, dr = \min\{\xi, \Xi_0\} \quad \text{for } \xi > 0. \tag{2.15}$$

We show that f is a travelling-wave solution of equation (1.1) on Ω with the properties we seek.

Plainly f is continuous in Ω. Furthermore, differentiating (2.14) and (2.15), $(a(f))'$ exists in (Ξ_1, Ξ_0) and satisfies (2.9). In fact, since

$$\theta(0) = 0, \quad \text{and,} \quad \theta(\delta) = 0 \quad \text{if } \Xi_1 > -\infty, \tag{2.16}$$

we deduce that $a(f) \in C^1(\Omega)$. Now define the mapping Ψ on $(0, \delta)$ by $\Psi(f(\xi)) = \xi$ for $\xi \in (\Xi_1, \Xi_0)$. By (2.14) and (2.15), Ψ is absolutely continuous on $(0, \delta)$ and $\Psi'(s) = -a'(s)/\theta(s)$ for almost all $s \in (0, \delta)$. Subsequently

$$\int_{\Psi(s_1)}^{\Psi(s_0)} |c(f(\xi))|\, d\xi = \int_{s_0}^{s_1} \frac{|c(r)|\, a'(r)}{\theta(r)}\, dr$$

for any $0 < s_0 < s_1 < \delta$. This implies $c \in L^1_{\text{loc}}(\Xi_1, \Xi_0)$, and thus $c_\varepsilon \in L^1_{\text{loc}}(\Omega)$ for small enough $\varepsilon > 0$, where c_ε is as defined in Section 2.1. Moreover, applying the above argument with c instead of $|c|$, (2.10) holds.

It follows that to complete the proof of Theorems 2.9, 2.10 and 2.11 it remains to establish (2.1). However, using (1.9), (2.9) and (2.10) there holds

$$((a(f))' + b(f) + \sigma f)(\xi_2) - ((a(f))' + b(f) + \sigma f)(\xi_1) = -\int_{\xi_1}^{\xi_2} c(f(\xi))\, d\xi$$

for any $\Xi_1 < \xi_1 < \xi_2 < \Xi_0$. This yields the absolute continuity of $(a(f))'+b(f)+\sigma f$ on (Ξ_1, Ξ_0) and that (2.8) holds almost everywhere on (Ξ_1, Ξ_0). So multiplying (2.8) by a test-function $\phi \in C_0^\infty(\Omega)$, integrating by parts and using (2.9) we compute

$$\begin{aligned}
&\int_{\xi_1}^{\xi_2} [\{(a(f))' + b(f) + \lambda f\}\phi' - c(f)\phi]\, d\xi \\
&\quad = \phi(\xi_2)\{-\theta(f(\xi_2)) + b(f(\xi_2)) + \lambda f(\xi_2)\} \\
&\qquad - \phi(\xi_1)\{-\theta(f(\xi_1)) + b(f(\xi_1)) + \lambda f(\xi_1)\}
\end{aligned} \tag{2.17}$$

for any $\Xi_1 < \xi_1 < \xi_2 < \Xi_0$. Letting $\xi_1 \downarrow \Xi_1$ and $\xi_2 \uparrow \Xi_0$ in (2.17) and recalling (2.16) yields (2.1).

2.4 Illustration of difficulties

Some aspects of the afore-going proof of Theorems 2.9, 2.10 and 2.11 may seem unnecessarily complicated. To illustrate the basic principles involved and some of the pitfalls which have been avoided, we discuss a number of specific examples below.

The exceptional feature of our first example is that in terms of the general notation it concerns an explicit wavefront solution from ℓ to 0 such that

$$f(\xi) = 0 \quad \text{for all } \xi \geq \xi^* \quad \text{some } \xi^* \in (-\infty, \infty) \tag{2.18}$$

and

$$f(\xi) = \ell \quad \text{for all } \xi \leq \xi^{**} \quad \text{some } \xi^{**} \in (-\infty, \infty). \tag{2.19}$$

This illustrates that the variables Ξ_0 and Ξ_1 introduced in the previous section, for both the proof of necessity and the proof of sufficiency in Theorems 2.9, 2.10 and 2.11, do not necessarily coincide with the supremum and infimum respectively of the domain of the travelling wave. It follows that, in general, Ξ_0 and Ξ_1 play an important role in the construction of a travelling-wave solution of equation (1.1) from a solution of the integral equation (1.9), and vice versa.

Application 2.21. *The equation*

$$u_t = (\sqrt{u(1-u)}u_x)_x + (k+2u)\sqrt{u(1-u)}, \tag{2.20}$$

where k is an arbitrary constant, admits the wavefront solution with wave speed $\sigma = k+1$ given by

$$f(\xi) = \begin{cases} 1 & \text{for } \xi \leq -\pi/2 \\ (1-\sin\xi)/2 & \text{for } -\pi/2 < \xi < \pi/2 \\ 0 & \text{for } \xi \geq \pi/2. \end{cases} \tag{2.21}$$

This equation falls into the general class (1.1) with $a(u) = \int_0^u \sqrt{s(1-s)}\,ds$, $b(u) = 0$, $c(u) = (k+2u)\sqrt{u(1-u)}$, and $I := [0,1]$. Differentiating (2.21), one may write

$$f'(\xi) = -\sqrt{f(1-f)}(\xi) \quad \text{for all } -\infty < \xi < \infty. \tag{2.22}$$

Subsequently, it is not too hard to check that (2.21) is a classical solution of the ordinary differential equation $(\sqrt{f(1-f)}f')' + (k+2f)\sqrt{f(1-f)} + \sigma f' = 0$ in $(-\infty,\infty)$, and, hence, a wavefront solution of (2.20) from 1 to 0 with wave speed σ. Now, in terms of the notation of the previous section, for this explicit travelling wave, one has $\omega = -\infty$, $\delta = 1$, $\Xi_0 := \sup\{\xi \in (\omega,\infty) : f(\xi) > 0\} = \pi/2$ and $\Xi_1 := \inf\{\xi \in (\omega,\infty) : f(\xi) < \delta\} = -\pi/2$. Hence, rewriting (2.9) as $\theta(f(\xi)) = -(a'(f)f')(\xi) = -(\sqrt{f(1-f)}f')(\xi)$ for $\xi \in (\Xi_1, \Xi_0)$, and using (2.22), produces the function

$$\theta(s) = s(1-s) \tag{2.23}$$

for $0 < s < 1$. This function is continuously extendible onto I with $\theta(0) = \theta(1) = 0$. It is easily verified that it satisfies the corresponding integral equation

$$\theta(s) = \sigma s - \int_0^s \frac{(k+2r)r(1-r)}{\theta(r)}\,dr \tag{2.24}$$

in this interval. Moreover, since it is positive on $(0,1)$, it automatically fulfils the integrability condition on I. Conversely, starting from the solution (2.23) of the integral equation (2.24), one may follow the methodology of the previous section to construct a travelling-wave solution of (2.20). To be specific, given any $\nu \in (0,1)$, using (2.13) – (2.15) one generates the function f on $(-\infty,\infty)$ given by

$$\arcsin\{1-2f(\xi)\} = \begin{cases} \max\{\xi - \arcsin(2\nu-1), -\pi/2\} & \text{for } \xi \le 0 \\ \min\{\xi - \arcsin(2\nu-1), \pi/2\} & \text{for } \xi > 0. \end{cases}$$

Modulo a translation, $\arcsin(2\nu - 1)$ with $0 < \nu < 1$, this gives the original wavefront solution (2.21).

Our next example is one of a wavefront solution that, excluding the possibility that (2.18) or (2.19) may hold, is not strictly monotonic. For such a travelling-wave, the measure of the set S defined in the previous section is positive. Thus, this complication cannot be ignored. Furthermore, we shall see that this can lead to a situation where the correspondence between distinct monotonic travelling-wave solutions of the reaction-convection-diffusion equation (1.1) and solutions of the integral equation (1.9) need not be one-to-one.

Application 2.22. *The equation*

$$u_t = u_{xx} + \begin{cases} 4u^2(8u-3) & \text{for } 0 \le u < 1/2 \\ 0 & \text{for } u = 1/2 \\ 4(1-u)^2(8u-5) & \text{for } 1/2 < u \le 1 \end{cases} \tag{2.25}$$

admits the wavefront solution with wave speed $\sigma = 0$ *given by*

$$f(\xi) = \begin{cases} [1+2(\xi_0-\xi)^2]/2[1+(\xi_0-\xi)^2] & \textit{for } \xi < \xi_0 \\ 1/2 & \textit{for } \xi_0 \le \xi \le \xi_1 \\ 1/2[1+(\xi-\xi_1)^2] & \textit{for } \xi > \xi_1 \end{cases} \tag{2.26}$$

for every pair of real numbers $\xi_0 < \xi_1$.

Proof. Differentiating (2.26), it can be checked that this function is globally continuously differentiable, and, twice continuously differentiable except at the points ξ_0 and ξ_1. Moreover, $u = f(x)$ satisfies (2.25) classically everywhere excepting for $x = \xi_0$ and $x = \xi_1$. Subsequently, it is easy to verify that this is a wavefront solution of equation (2.25) from 1 to 0 with wave speed 0 in the sense of Definition 2.2. Following the analysis of the previous section, we find that $\Xi_0 := \infty$ and $\Xi_1 := -\infty$; while the function θ can be defined by

$$\theta(s) = \begin{cases} 2s\sqrt{2s(1-2s)} & \text{for } 0 \le s \le 1/2 \\ 2(1-s)\sqrt{2(1-s)(2s-1)} & \text{for } 1/2 < s \le 1. \end{cases} \tag{2.27}$$

This function may be verified to satisfy the corresponding integral equation (1.9). The set $S = [\xi_0, \xi_1]$. Note that $c(f(\xi)) = 0$ for all $\xi \in S$, in conformance with Lemma 2.20. Conversely, we may follow the analysis of the previous section to construct a travelling-wave solution f of equation (2.25) from the function (2.27). Given any $\nu \in (0,1)$, using (2.13) – (2.15), we obtain

$$f(\xi) = \begin{cases} [1+2(\eta-\xi)^2]/2[1+(\eta-\xi)^2] & \text{for } \xi \le \eta \\ 1/2[1+(\xi-\eta)^2] & \text{for } \xi > \eta, \end{cases} \tag{2.28}$$

where

$$\eta = \begin{cases} -\sqrt{(1-2\nu)/2\nu} & \text{for } \nu \le 1/2 \\ \sqrt{(2\nu-1)/2(1-\nu)} & \text{for } \nu > 1/2. \end{cases} \tag{2.29}$$

This is the wavefront solution (2.26) with $\xi_0 = \xi_1 = \eta$. □

Let us recapitulate the above. Starting with an explicit travelling-wave solution (2.26) of the partial differential equation (2.25), we computed a solution (2.27) of the appropriate integral equation (1.9). Thereafter, using (2.27), we constructed a travelling-wave solution (2.28), (2.29) of the partial differential equation (2.25). During this process though, more than a translation of the wave has occurred. Although the original solution is constant in an interval $[\xi_0, \xi_1]$ of positive length, the reconstructed solution (2.28), (2.29) does not possess this property. In fact, the reconstructed solution is independent of the magnitude of $\xi_1 - \xi_0$. Thus, in general, there is not a one-to-one correspondence between distinct monotonic travelling-wave solutions of the reaction-convection-diffusion equation (1.1) and solutions of the integral equation (1.9). Notwithstanding, from the analysis in the previous section, it can be discerned that there is a one-to-one correspondence under the proviso that one considers only those monotonic travelling-wave solutions

f defined in a domain Ω, with $\inf\{f(\xi) : \xi \in \Omega\} = 0$ and $\sup\{f(\xi) : \xi \in \Omega\} = \delta$ say, which are strictly monotonic in $\{\xi \in \Omega : 0 < f(\xi) < \delta\}$.

Our final example displays a further complexity. This is that the set $S := \{\xi \in (\Xi_1, \Xi_0) : (a(f))'(\xi) = 0\}$ has an infinite number of disjoint components. Moreover, it is possible to define the measure of these components in a such a way that (2.18) may or may not occur. In fact, we could extend this example to control the occurrence of (2.19) as well. We spare the reader this procedure.

Application 2.23. *Let $\{s_i\}_{i=0}^{\infty}$ be a strictly decreasing sequence of values in $(0,1]$ such that $s_0 = 1$ and $s_i = \mathcal{O}(i^{-3})$ as $i \to \infty$. Set $\xi_0 = -\pi$. Thereafter, by induction, define $\eta_{i-1} = \xi_{i-1} + \pi/i^2$ and $\xi_i > \eta_{i-1}$ for all $i \geq 1$. The equation*

$$u_t = u_{xx} + \begin{cases} i^4\left(u - \frac{s_i + s_{i-1}}{2}\right) & \textit{for } s_i < u < s_{i-1} \textit{ and } i \geq 1 \\ 0 & \textit{otherwise} \end{cases}$$

admits the wavefront solution with wave speed $\sigma = 0$ given by

$$f(\xi) = \begin{cases} 1 & \textit{for } \xi \leq \xi_0 \\ \frac{s_i + s_{i-1}}{2} - \frac{s_i - s_{i-1}}{2}\cos\{i^2(\xi - \xi_{i-1})\} & \textit{for } \xi_{i-1} < \xi < \eta_{i-1} \textit{ and } i \geq 1 \\ s_i & \textit{for } \eta_{i-1} \leq \xi \leq \xi_i \textit{ and } i \geq 1 \\ 0 & \textit{otherwise.} \end{cases} \tag{2.30}$$

Proof. It requires some laborious calculation to verify this example, and, details will be omitted. The corresponding solution of the integral equation (1.9) is

$$\theta(s) = \begin{cases} i^2\sqrt{(s - s_i)(s_{i-1} - s)} & \text{if } s_i < s < s_{i-1} \text{ and } i \geq 1 \\ 0 & \text{otherwise.} \end{cases} \tag{2.31}$$

The crucial features of this example are that

$$\begin{aligned} \Xi_0 &:= \sup\{\xi \in (-\infty, \infty) : f(\xi) > 0\} \\ &= \sum_{i=1}^{\infty}(\eta_i - \xi_i) + \sum_{i=1}^{\infty}(\xi_i - \eta_{i-1}) \\ &= \sum_{i=1}^{\infty}\frac{\pi}{(i+1)^2} + \sum_{i=1}^{\infty}(\xi_i - \eta_{i-1}) \end{aligned} \tag{2.32}$$

and $S = \bigcup_{i=1}^{\infty}[\eta_{i-1}, \xi_i]$. Thus, (2.18) holds if and only if the final sum in (2.32) converges. On the other hand, the construction of a travelling-wave solution from (2.31) following the approach presented in the previous section automatically results, modulo translation, in the function (2.30) with $\xi_i = \eta_{i-1}$ for all $i \geq 1$. □

Both this and the preceding example concern a wavefront solution f from ℓ to 0 with the property that there exists one or more values $s^* \in (0, \ell)$ such that $f \equiv s^*$ in some interval of positive length. We have seen too that deriving a solution θ of the integral equation from this wavefront solution and then using the methodology outlined in the previous section to reconstruct a travelling-wave solution results in a wavefront solution that does not have this property. However, this need not be the case. Noting, that necessarily $\theta(s^*) = c(s^*) = 0$, it is possible to insert an interval of arbitrary length on which $f \equiv s^*$ in the reconstructed solution. To be more specific, given any solution θ of the integral equation (1.9), defined in an interval $[0, \delta)$, or, $[0, \delta]$ with $\delta < \infty$ say, for a countable number of values $s^* \in (0, \delta)$ such that $\theta(s^*) = 0$, in the light of the lemma below one can modify the proof in the previous section to construct a decreasing travelling-wave solution f of equation (1.1) in a domain Ω such that $\inf\{f(\xi) : \xi \in \Omega\} = 0$, $\sup\{f(\xi) : \xi \in \Omega\} = \delta$, and, $f \equiv s^*$ in a subinterval of Ω with positive length. This device will be employed in the proof of Corollary 2.30 of Theorem 2.9, Corollary 2.34 of Theorem 2.10, and Corollary 2.38 of Theorem 2.11, below.

Lemma 2.24. *Let θ be a solution of equation* (1.9) *on* $[0, \delta)$ *for some* $0 < \delta \leq \ell$. *Suppose that* $\theta(s^*) = 0$ *for some* $0 < s^* < \delta$. *Then* $c(s^*) = 0$.

This result on solutions of the integral equation is equivalent to parts (iii) and (iv) of Lemma 2.20 on travelling-wave solutions of the partial differential equation (1.1).

Proof of Lemma 2.24. By equation (1.9) there holds

$$\theta(s) = \sigma s + b(s) - \sigma s^* - b(s^*) - \int_{s^*}^{s} \frac{c(r)a'(r)}{\theta(r)}\, dr \quad \text{for all } 0 < s < \delta. \tag{2.33}$$

Subsequently, $\widetilde{\theta}(s) := \theta(s + s^*)$ satisfies the equation

$$\widetilde{\theta}(s) = \sigma s + \widetilde{b}(s) - \int_{0}^{s} \frac{\widetilde{c}(r)\widetilde{a}'(r)}{\widetilde{\theta}(r)}\, dr$$

with $\widetilde{b}(s) := b(s + s^*) - b(s^*)$, $\widetilde{c}(s) := c(s + s^*)$ and $\widetilde{a}(s) := a(s + s^*) - a(s^*)$ for $s \in [0, \delta - s^*)$. However, by Hypothesis 2.1 and Lemma 2.15(ii), this integral equation has no solution if $c(s^*) > 0$. Likewise, $\widetilde{\theta}(s) := \theta(s^* - s)$ satisfies

$$\widetilde{\theta}(s) = -\sigma s + \widetilde{b}(s) - \int_{0}^{s} \frac{\widetilde{c}(r)\widetilde{a}'(r)}{\widetilde{\theta}(r)}\, dr$$

with $\widetilde{b}(s) := b(s^* - s) - b(s^*)$, $\widetilde{c}(s) := -c(s^* - s)$ and $\widetilde{a}(s) := a(s^*) - a(s^* - s)$ for $s \in [0, s^*)$, and, by Hypothesis 2.1 and Lemma 2.15(ii) this integral equation has no solution for $c(s^*) < 0$. □

To close this section, we draw the following conclusions from the discussion surrounding the examples. This will be useful for later.

Lemma 2.25. *To every solution θ of the integral equation* (1.9) *satisfying the integrability condition in an interval* $[0,\delta)$ *for some* $0<\delta\leq\ell$ *there corresponds precisely one distinct semi-wavefront solution f of equation* (1.1) *with wave speed σ decreasing to* 0 *defined in a domain Ω such that* $\sup\{f(\xi):\xi\in\Omega\}=\delta$ *and f is strictly monotonic in* $\{\xi\in\Omega:\delta>f(\xi)>0\}$. *Moreover, there corresponds no other distinct semi-wavefront solution f of equation* (1.1) *with wave speed σ decreasing to* 0 *defined in a domain Ω such that* $\sup\{f(\xi):\xi\in\Omega\}=\delta$, *if and only if* $\theta(s)>0$ *for all* $0<s<\delta$.

Lemma 2.26. *Suppose that* $\ell<\infty$. *To every solution θ of the integral equation* (1.9) *satisfying the integrability condition in* $[0,\ell]$ *and* $\theta(\ell)=0$ *there corresponds precisely one distinct wavefront solution f of equation* (1.1) *from ℓ to* 0 *with wave speed σ such that f is strictly monotonic in* $\{\xi\in(-\infty,\infty):\ell>f(\xi)>0\}$. *Moreover, there corresponds no other distinct wavefront solution of equation* (1.1) *from ℓ to* 0 *with wave speed σ if and only if* $\theta(s)>0$ *for all* $0<s<\ell$.

2.5 Further properties

The following are corollaries of the above proof of Theorems 2.9, 2.10 and 2.11 that are of particular relevance to Theorem 2.9. The last three are illuminated by the examples in the previous section and the discussion surrounding them.

Corollary 2.27 (of Theorem 2.9). *If f is a semi-wavefront solution of* (1.1) *decreasing to* 0 *in an interval Ω, then* $a(f)\in C^1(\Omega)$ *and* $(a(f))'(\xi)\to 0$ *as* $\xi\to\infty$.

Corollary 2.28 (of Theorem 2.9). *If equation* (1.1) *has a semi-wavefront solution with speed σ decreasing to* 0, *then the equation has at least one such solution f in an interval Ω with the property that f is strictly decreasing in* $\{\xi\in\Omega:f(\xi)>0\}$.

Corollary 2.29 (of Theorem 2.9). *Equation* (1.1) *has none, one or an infinite number of distinct semi-wavefront solutions f with speed σ decreasing to* 0 *in an interval Ω which are strictly decreasing in* $\{\xi\in\Omega:f(\xi)>0\}$ *according to whether equation* (1.9) *has respectively none, one, or an infinite number of solutions satisfying the integrability condition on an interval* $[0,\delta)$ *with* $0<\delta\leq\ell$.

Corollary 2.30 (of Theorem 2.9). (i) *Equation* (1.1) *has a semi-wavefront solution f with speed σ in an interval Ω for which*

$$f(\xi)=0 \quad \textit{for all } \xi\geq\xi^* \quad \textit{some } \xi^*\in\Omega \tag{2.34}$$

if and only if equation (1.9) *has a solution θ on an interval* $[0,\delta)$ *with the property*

$$\int_0^s \frac{a'(r)}{\theta(r)}\,dr<\infty \quad \textit{for all } 0<s<\delta. \tag{2.35}$$

(ii) *Equation* (1.1) *has a semi-wavefront solution* f *with speed* σ *decreasing to* 0 *in an interval* Ω *with*

$$f(\xi) > 0 \quad \text{for all } \xi \in \Omega \tag{2.36}$$

if and only if (1.9) *has a solution* θ *satisfying the integrability condition on an interval* $[0, \delta)$ *with the property*

$$\int_0^s \frac{a'(r)}{\theta(r)}\, dr = \infty \quad \text{for all } 0 < s < \delta \tag{2.37}$$

or $\theta(s_i) = 0$ *for a sequence of values* $\{s_i\}_{i=1}^{\infty} \subset (0, \delta)$ *such that* $s_i \downarrow 0$ *as* $i \uparrow \infty$.

Proof. (i) If equation (1.1) has a travelling-wave solution with the characteristics described, then, in the proof of the 'only if' part of Theorem 2.9 in Section 2.3, necessarily $\Xi_0 < \infty$. Subsequently letting $\xi_2 \uparrow \Xi_0$ in (2.11) yields (2.35) for the solution θ of equation (1.9) on $[0, \delta)$. On the other hand, if (1.9) has a solution θ satisfying (2.35), then our constructive proof of the 'if' part of Theorem 2.9 in Section 2.3 yields a travelling-wave solution f on $(0, \infty)$ for which (2.34) holds.

(ii) If equation (1.1) has a travelling-wave solution with the property now described, then in Section 2.3 necessarily $\Xi_0 = \infty$. In this case, letting $\xi_2 \uparrow \Xi_0$ in (2.11) implies that (2.37) holds for the solution θ of (1.9) on $[0, \delta)$, or, $(\xi_1, \infty) \cap S = \{\xi \in (\xi_1, \infty) : \theta(f(\xi)) = 0\}$ has infinite measure for every $\xi_1 > \Xi_1$. However, since $\theta(f(\xi)) = 0$ for all $\xi \in S$, the latter can only be the case if there is a sequence $\{s_i\}_{i=1}^{\infty} \subset (0, \delta)$ such that $s_i \downarrow 0$ as $i \uparrow \infty$ and $\theta(s_i) = 0$ for all $i \geq 1$. This proves the necessity. Conversely, if the integral equation (1.9) has a solution θ satisfying the integrability condition on $[0, \delta)$ for which (2.37) holds, then our constructive proof of the 'if' part of Theorem 2.9 directly yields a travelling-wave solution f in a domain Ω with f satisfying (2.36). Whereas, if this is not the case, but (1.9) admits a solution θ on an interval $[0, \delta)$ with a sequence $\{s_i\}_{i=1}^{\infty}$ as described, then we can modify our construction slightly as intimated in Application 2.23 and its subsequent discussion. We set $\nu = s_1$, $\eta_0 := 0$ and thereafter by induction define $\xi_i := \eta_{i-1} + 1$ and

$$\eta_i := \xi_i + \int_{s_{i+1}}^{s_i} \frac{a'(r)}{\theta(r)}\, dr$$

for all $i \geq 1$. We let f be given by (2.14) where Ξ_1 is defined as of old by (2.13), $f(\xi) = s_i$ for $\eta_{i-1} \leq \xi \leq \xi_i$ and

$$\int_{f(\xi)}^{s_i} \frac{a'(r)}{\theta(r)}\, dr = \xi - \xi_i \quad \text{for } \xi_i < \xi < \eta_i$$

for each $i \geq 1$. Following the earlier proof in Section 2.3 and recalling Lemma 2.24, this can be shown to still yield a travelling-wave solution of equation (1.1). Moreover this solution is such that (2.36) holds. □

Similarly we can prove the following corollaries to Theorem 2.10. Throughout implicitly $\ell < \infty$.

Corollary 2.31 (of Theorem 2.10). *If f is a wavefront solution of* (1.1) *from ℓ to 0, then $a(f) \in C^1(-\infty,\infty)$ and $(a(f))'(\xi) \to 0$ as $\xi \to \pm\infty$.*

Corollary 2.32 (of Theorem 2.10). *If equation* (1.1) *has a wavefront solution from ℓ to 0 with speed σ, then the equation has at least one such solution f with the property that f is strictly decreasing in $\{\xi \in (-\infty,\infty) : \ell > f(\xi) > 0\}$.*

Corollary 2.33 (of Theorem 2.10). *Equation* (1.1) *has none, one or an infinite number of distinct wavefront solutions f from ℓ to 0 with speed σ which are strictly decreasing in $\{\xi \in (-\infty,\infty) : \ell > f(\xi) > 0\}$ according to whether equation* (1.9) *has respectively none, one, or an infinite number of solutions θ satisfying the integrability condition on $[0,\ell]$ with $\theta(\ell) = 0$.*

Corollary 2.34 (of Theorem 2.10). (i) *Equation* (1.1) *has a wavefront solution f with speed σ for which* (2.18) *and* (2.19) *hold if and only if* (1.9) *has a solution θ on $[0,\ell]$ with the properties $\theta(\ell) = 0$ and*

$$\int_0^\ell \frac{a'(r)}{\theta(r)}\,dr < \infty. \tag{2.38}$$

(ii) *Equation* (1.1) *has a wavefront solution f from ℓ to 0 with speed σ for which* (2.18) *holds but*

$$f(\xi) < \ell \quad \text{for all } \xi \in (-\infty,\infty) \tag{2.39}$$

if and only if (1.9) *has a solution θ on $[0,\ell]$ with the properties*

$$\int_0^s \frac{a'(r)}{\theta(r)}\,dr < \infty \quad \text{for all } 0 < s < \ell, \tag{2.40}$$

and,

$$\int_0^\ell \frac{a'(r)}{\theta(r)}\,dr = \infty \tag{2.41}$$

or $\theta(s_i) = 0$ for a sequence of values $\{s_i\}_{i=1}^\infty \subset (0,\ell)$ such that $s_i \uparrow \ell$ as $i \uparrow \infty$.

(iii) *Equation* (1.1) *has a wavefront solution f from ℓ to 0 with speed σ for which*

$$f(\xi) > 0 \quad \text{for all } \xi \in (-\infty,\infty) \tag{2.42}$$

and (2.19) *holds if and only if* (1.9) *has a solution θ on $[0,\ell]$ with the properties $\theta(\ell) = 0$,*

$$\int_s^\ell \frac{a'(r)}{\theta(r)}\,dr < \infty \quad \text{for all } 0 < s < \ell, \tag{2.43}$$

and, (2.41) *is satisfied or $\theta(s_i) = 0$ for a sequence of values $\{s_i\}_{i=1}^\infty \subset (0,\ell)$ such that $s_i \downarrow 0$ as $i \uparrow \infty$.*

(iv) *Equation* (1.1) *has wavefront solution f from ℓ to 0 with speed σ for which* (2.39) *and* (2.42) *hold if and only if* (1.9) *has a solution θ satisfying the integrability condition on* $[0,\ell]$ *with the properties*

$$\int_0^s \frac{a'(r)}{\theta(r)}\,dr = \infty \quad \text{for all } 0 < s < \ell \tag{2.44}$$

or $\theta(s_i) = 0$ *for a sequence of values* $\{s_i\}_{i=1}^{\infty} \subset (0,\ell)$ *such that* $s_i \downarrow 0$ *as* $i \uparrow \infty$, *and,*

$$\int_s^\ell \frac{a'(r)}{\theta(r)}\,dr = \infty \quad \text{for all } 0 < s < \ell \tag{2.45}$$

or $\theta(s_i) = 0$ *for a sequence of values* $\{s_i\}_{i=1}^{\infty} \subset (0,\ell)$ *such that* $s_i \uparrow \ell$ *as* $i \uparrow \infty$.

Analogous results for Theorem 2.11 are the following. In these results it is assumed that $\ell = \infty$.

Corollary 2.35 (of Theorem 2.11). *If f is an unbounded monotonic travelling-wave solution of* (1.1) *in an interval* $\Omega = (\omega,\infty)$ *with* $-\infty \leq \omega < \infty$ *decreasing to* 0 *then* $a(f) \in C^1(\Omega)$ *and* $(a(f))'(\xi) \to 0$ *as* $\xi \to \infty$.

Corollary 2.36 (of Theorem 2.11). *If equation* (1.1) *has an unbounded monotonic travelling-wave solution with speed σ decreasing to* 0, *then the equation has at least one such solution f in an interval Ω with the property that f is strictly decreasing in* $\{\xi \in \Omega : f(\xi) > 0\}$.

Corollary 2.37 (of Theorem 2.11). *Equation* (1.1) *has none, one or an infinite number of distinct unbounded monotonic travelling-wave solutions f with speed σ in a domain* $\Omega = (\omega,\infty)$ *with* $-\infty \leq \omega < \infty$ *decreasing to* 0 *which are strictly decreasing in* $\{\xi \in \Omega : f(\xi) > 0\}$ *according to whether equation* (1.9) *has respectively none, one, or an infinite number of solutions satisfying the integrability condition on* $[0,\infty)$.

Corollary 2.38 (of Theorem 2.11). (i) *Equation* (1.1) *has an unbounded strict semi-wavefront solution f with speed σ for which* (2.34) *holds if and only if* (1.9) *has a solution θ on* $[0,\infty)$ *with the property that* (2.38) *holds.*

(ii) *Equation* (1.1) *has an unbounded monotonic global travelling-wave solution f with speed σ for which* (2.34) *holds if and only if* (1.9) *has a solution θ on* $[0,\infty)$ *with the properties that* (2.40) *holds, and,* (2.41) *holds or* $\theta(s_i) = 0$ *for a sequence of values* $\{s_i\}_{i=1}^{\infty} \subset (0,\infty)$ *such that* $s_i \uparrow \infty$ *as* $i \uparrow \infty$.

(iii) *Equation* (1.1) *has an unbounded strict semi-wavefront solution f with speed σ decreasing to* 0 *for which* (2.36) *holds if and only if* (1.9) *has a solution θ on* $[0,\infty)$ *with the properties that* (2.43) *holds, and,* (2.41) *holds or* $\theta(s_i) = 0$ *for a sequence of values* $\{s_i\}_{i=1}^{\infty} \subset (0,\infty)$ *such that* $s_i \downarrow 0$ *as* $i \uparrow \infty$.

(iv) *Equation* (1.1) *has an unbounded monotonic global travelling-wave solution* f *with speed* σ *decreasing to* 0 *such that* (2.36) *holds if and only if* (1.9) *has a solution* θ *satisfying the integrability condition on* $[0, \infty)$ *with the properties that* (2.44) *holds or* $\theta(s_i) = 0$ *for a sequence of values* $\{s_i\}_{i=1}^{\infty} \subset (0, \infty)$ *such that* $s_i \downarrow 0$ *as* $i \uparrow \infty$, *and,* (2.45) *holds or* $\theta(s_i) = 0$ *for a sequence of values* $\{s_i\}_{i=1}^{\infty} \subset (0, \infty)$ *such that* $s_i \uparrow \infty$ *as* $i \uparrow \infty$.

2.6 Classical results

In Section 2.1, in mitigation of the consideration of only monotonic travelling waves, it was mentioned that under certain circumstances any global travelling-wave solution f of an equation of the type (1.1) such that (2.2) and (2.3) hold for some $\ell^-, \ell^+ \in I$ with $c(\ell^-) = c(\ell^+) = 0$ and such that the values of f lie in the range $\min\{\ell^-, \ell^+\} \leq f \leq \max\{\ell^-, \ell^+\}$ is necessarily monotonic. The justification of this remark can be made without any loss of generality under the assumption that $\ell^- = \ell$ and $\ell^+ = 0$. Extending arguments in [103] the following is then the result which can be proven.

Theorem 2.39. *Suppose that* $a \in C^2(I)$ *with* $a'(u) > 0$ *for all* $u \in I$, $b \in C^1(I)$ *and* $c \in C^1(I)$. *Let* f *be a global travelling-wave solution of equation* (1.1) *such that* $f(\xi) \to \ell$ *as* $\xi \to -\infty$ *and* $f(\xi) \to 0$ *as* $\xi \to \infty$. *Then* $f \in C^2(-\infty, \infty)$, *and,* $(a(f))'(\xi) < 0$ *for all* $-\infty < \xi < \infty$.

Proof. The key to this result is the observation that if $c \in C(I)$, then it can be verified that $c(f) \in L^1_{\text{loc}}(\Omega)$ for any travelling-wave solution f of equation (1.1) in a domain Ω in the sense of Definition 2.2. Subsequently by the regularity argument behind part (ii) of Lemma 2.20 it can be demonstrated that any such travelling wave must be a classical solution of the ordinary differential equation (1.3). The remaining conclusions follow from analysis of the system (1.7). □

A number of conclusions about solutions of the integral equation (1.9) may be drawn from the above result. However for completeness and because of its usefulness later as a more general conclusion, we prove the following studying the integral equation only.

Lemma 2.40. *Suppose that* ca' *is differentiable in* $(0, \delta)$ *for some* $0 < \delta \leq \ell$. *Then any solution* θ *of equation* (1.9) *satisfying the integrability condition on* $[0, \delta)$ *is necessarily positive on* $(0, \delta)$.

Conversely of course, any solution θ of equation (1.9) that is positive in an interval $(0, \delta)$ automatically satisfies the integrability condition. Thus it follows from Lemma 2.40 that if ca' is differentiable in $(0, \ell)$, the search for a solution of the integral equation which satisfies the integrability condition can be reduced to the search for a positive solution.

Proof of Lemma 2.40. Suppose that there exists a point $s^* \in (0,\delta)$ such that $\theta(s^*) = 0$. Then, by Lemma 2.24, $(ca')(s^*) = 0$. Furthermore, arguing as in the proof of Lemma 2.24, (2.33) holds. Therefore, for any $\beta > |\sigma + b'(s^*)|$ and $\gamma > |(ca')'(s^*)|$ we can choose an $\iota < \min\{s^*, \delta - s^*\}$ so small that

$$\theta(s) \leq \left\{\beta + \gamma \int_{s^*-\iota}^{s^*+\iota} \frac{1}{\theta(r)}\, dr\right\} |s - s^*| \tag{2.46}$$

for all $s^* - \iota < s < s^* + \iota$. On the other hand, because $a \in C^1(0,\ell)$ and $a' > 0$ in $(0,\ell)$, the integrability condition implies that

$$\int_{s^*-\iota}^{s^*+\iota} \frac{1}{\theta(r)}\, dr < \infty. \tag{2.47}$$

Together, (2.46) and (2.47) are incompatible. Thus the lemma is proven by *reductio ad absurdum*. □

Another point made in Section 2.1, in mitigation of the concentration on wavefront solutions, is that under appropriate conditions the wavefront solutions are the only bounded monotonic global travelling-wave solutions of an equation of the class (1.1). This is borne out by the following.

Theorem 2.41. *Suppose that* $c \in C(0,\ell)$. *Suppose furthermore that equation* (1.1) *has a bounded monotonic global travelling-wave solution* f. *Let* $\ell^-, \ell^+ \in I$ *be such that* (2.2) *and* (2.3) *hold. Then* $c(\ell^-) = c(\ell^+) = 0$.

Proof. It suffices to prove that $K := c(\ell^+) = 0$, since the class of equations (1.1) is invariant under the change of variable $x \mapsto -x$. To do this, suppose for the sake of argument that $K \neq 0$. Then, $0 < \ell^+ < \ell$, and, by Lemma 2.20(i), $f(\xi) \neq \ell^+$ for all $-\infty < \xi < \infty$. Subsequently, by an affine transformation of the dependent variable, we involve no loss of generality if we suppose that f is a semi-wavefront solution decreasing to 0 for which (2.36) holds, for an alternative equation of the class (1.1) with the property that $c(u) \to K$ as $u \downarrow 0$. However, in this case, letting σ denote the wave speed of f, by Corollary 2.30(ii), there exists a $\delta > 0$ such that equation (1.9) has a solution θ on $[0,\delta)$ satisfying (2.37) or $c(s_i) = 0$ for a sequence of values $\{s_i\}_{i=1}^{\infty} \subset (0,\delta)$ such that $s_i \to 0$ as $i \to \infty$. Simultaneously, since $c(s) \to K$ as $s \downarrow 0$, the first of these deductions is incompatible with the finiteness of the integral in (1.9) for every $0 < s < \delta$, while the second is excluded a priori. Thus, from the supposition that $K \neq 0$, we arrive at a contradiction. We have to concede that $K = 0$. □

Corollary 2.42 (of Theorem 2.41). *Suppose that* $\ell = \infty$, $c \in C(0,\infty)$, *and,* $c(u) \neq 0$ *for all* $u > 0$. *Then, beside the constant solution* $f \equiv 0$, *equation* (1.1) *has no bounded monotonic global travelling-wave solutions. Furthermore, besides possible monotonic semi-wavefront solutions decreasing to* 0, *the equation has no unbounded nonincreasing travelling-wave solutions.*

Bibliographic notes

Necessarily the introduction does not include all areas in which partial differential equations of the class (1.1) arise. General surveys covering diverse fields of application are [10, 11, 41, 74, 75, 77, 82, 96, 101, 107, 134, 221, 230, 273]. Different fields with suitable references for specific prototypes include: heat transfer [78,316,317], combustion [31,58,59,106,314,315], reaction chemistry [13,14], fluid dynamics [57], plasma physics [46,66], soil-moisture physics [30,68,238,278], foam drainage [293, 304], crystal growth [139, 279], biological population genetics [56,162,170,208,210,211,222], cellular ecology [269], neurology [266,267], and, synergy [152].

The role which has been played by travelling-wave solutions in the development of the theory of nonlinear reaction-convection-diffusion equations can be discerned from [8,9,72,78,80,81,111,112,114,119,126,127,129,161,185,223,228]. Travelling waves are of course a special kind of self-similar solution, and, the role which these solutions play in intermediate asymptotic behaviour in general is discussed in [24,27,28] among other works.

The statement that Hypothesis 2.1 is on the borderline of the current theory for equations of the class (1.1) is justified in [113, 128]. In [113], the effect of a convection term which falls outside the hypothesis is discussed. In [128], the same is done for a reaction term.

The term *wavefront* for a monotonic global travelling-wave solution which connects two equilibrium states of an evolution equation is commonplace at the present time; due, among other reasons, no doubt, to its coinage in [56,98,101,211].

We could view the relationship between travelling-wave solutions of equations of the type (1.1) and the ordinary differential equation (1.8) as a step towards our integral equation approach. This has been noted and utilized in the past by various authors [20–23, 32–38, 47, 64, 98, 101, 103, 104, 173, 176, 186, 197, 210, 211, 232, 241, 276, 277, 280, 284, 295, 297, 299, 314]. For a related problem, see [166, 167, 314].

Actually, the equivalence between travelling-wave solutions of equation (1.1) and the integral equation (1.9) demonstrated in Section 2.3 may be shown under weaker conditions on the coefficients a, b and c than those in Hypothesis 2.1. In particular, equivalences can also be drawn in the cases $c(0) \neq 0$, and, $c(\ell) \neq 0$ when $\ell < \infty$. We refer to [124] for such and other details.

Travelling-wave solutions of nonlinear second-order hyperbolic equations similar to those of the nonlinear second-order parabolic equation (1.1) have been investigated in [78, 146–148]; while the analogous solutions of first-order nonlinear conservation laws have been studied in [200].

Very much bound up with the study of the occurrence of travelling-wave solutions of equations of the type (1.1) is the study of the stability of such waves.

This is a topic in its own right, and, far too broad and fragmentary to be covered in this monograph. It is investigated in more detail for various variants of (1.1) in [40,49,51,53–55,79,82,85,95,99,101–104,115,134,135,149,151,162–165,168,169, 173–176,184,186,190,191,202,203,207,224,234,241,249,251,252,262,263,276,277, 283,284,286,288,290,291,297,298]. Worthy of special mention for an introduction in historical perspective are the works of Kolmogorov, Petrovskii and Piskunov [186], of Kanel' [174–176], of McKean [201,202], of Aronson and Weinberger [20,21], of Fife and McLeod [102–104], of Sattinger [262,263], of Uchiyama [283–286], and of Bramson [53–55].

Chapter 3

Transformations

There are a number of transformations known, through which the existence of a wavefront solution of one reaction-convection-diffusion equation can be used to determine the existence of a wavefront solution of another. These transformations become quite transparent in terms of the integral equation (1.9). In fact, they are embodied in the following two theorems.

Theorem 3.1. *Suppose that $\ell < \infty$. If either one of the equations*

$$u_t = (a(u))_{xx} + (b(u))_x + c(u) \tag{3.1}$$

or

$$u_t = u_{xx} + (b(u))_x + c(u)a'(u) \tag{3.2}$$

admits a wavefront solution from ℓ to 0 with speed σ, they both do. Moreover if $a \in C^1(I)$, it is possible to define such a solution f_1 of the first equation and such a solution f_2 of the second equation, for which $f_2(\xi) = f_1(\Psi(\xi))$ for all $\xi \in (-\infty, \infty)$, where

$$\Psi(\xi) = \int_0^\xi a'(f_2(\eta))\, d\eta. \tag{3.3}$$

Proof. The integral equations (1.9) associated with the differential equations (3.1) and (3.2) are identical. So the assertions regarding the existence of the travelling waves are an immediate corollary of Theorem 2.10. Furthermore, if θ denotes the appropriate solution of the integral equation (1.9), for any $0 < \nu < \ell$, a wavefront solution f_1 of equation (3.1) can be constructed by

$$\int_{f_1(\xi)}^{\nu} \frac{a'(s)}{\theta(s)}\, ds = \xi \quad \text{for } \Xi_1^{(1)} < \xi < \Xi_0^{(1)}, \tag{3.4}$$

where

$$\Xi_1^{(1)} := -\int_\nu^\ell \frac{a'(s)}{\theta(s)}\, ds \quad \text{and} \quad \Xi_0^{(1)} := \int_0^\nu \frac{a'(s)}{\theta(s)}\, ds,$$

and a wavefront solution f_2 of equation (3.2) via

$$\int_{f_2(\xi)}^{\nu} \frac{1}{\theta(s)}\, ds = \xi \quad \text{for } \Xi_1^{(2)} < \xi < \Xi_0^{(2)}, \tag{3.5}$$

where

$$\Xi_1^{(2)} := -\int_\nu^\ell \frac{1}{\theta(s)}\, ds \quad \text{and} \quad \Xi_0^{(2)} := \int_0^\nu \frac{1}{\theta(s)}\, ds.$$

Whence the travelling-wave solutions of the respective equations can be related by $f_2(\xi) = f_1(\Psi(\xi))$ for any $\Xi_1^{(2)} < \xi < \Xi_0^{(2)}$ for some transformation function Ψ. Using (3.4) and (3.5) this transformation function can be identified as

$$\begin{aligned}\Psi(\xi) &= \int_{f_1(\Psi(\xi))}^{\nu} \frac{a'(s)}{\theta(s)}\,ds \\ &= \int_{f_2(\xi)}^{\nu} \frac{a'(s)}{\theta(s)}\,ds \\ &= -\int_0^{\xi} \frac{a'(f_2(\eta))}{\theta(f_2(\eta))} f_2'(\eta)\,d\eta \\ &= \int_0^{\xi} a'(f_2(\eta))\,d\eta.\end{aligned}$$

Moreover, since when $a \in C^1(I)$, $\Xi_0^{(2)} < \infty$ automatically infers $\Xi_0^{(1)} < \infty$ and likewise $\Xi_1^{(2)} > -\infty$ implies $\Xi_1^{(1)} > -\infty$, this transformation can be extended to $-\infty < \xi < \infty$. □

Theorem 3.2. *Suppose that $\ell < \infty$. Let $b \in C^1(0, \ell)$ and f be a wavefront solution of the equation*

$$u_t = u_{xx} + (b(u))_x \tag{3.6}$$

from ℓ to 0 with speed σ. Then f is similarly a wavefront solution of the equation

$$u_t = u_{xx} - b'(u)\{\sigma u + b(u)\}. \tag{3.7}$$

Conversely, let $c \in C^1(0, \ell)$ and f be a wavefront solution of the equation

$$u_t = u_{xx} + c(u) \tag{3.8}$$

from ℓ to 0 with speed σ. Then f is similarly a wavefront solution of the equation (3.6) for some function b such that

$$-b'(u)\{\sigma u + b(u)\} = c(u). \tag{3.9}$$

Proof. The integral equation associated with (3.6) is simply $\theta(s) = \sigma s + b(s)$. Subsequently, if this function is nonnegative it trivially satisfies the integral equation associated with (3.7). The converse case is a little more subtle. If (3.8) has a travelling wave of the stated type, then following Theorem 2.10 the integral equation

$$\theta(s) = \sigma s - \int_0^s \frac{c(r)}{\theta(r)}\,dr \tag{3.10}$$

has a solution θ satisfying the integrability condition on $[0, \ell]$ with $\theta(\ell) = 0$.

Subsequently defining $b(s) := \theta(s) - \sigma s$ we have

$$-b(s) = \int_0^s \frac{c(r)}{\sigma r + b(r)}\, dr$$

which yields (3.9). Simultaneously, θ is a 'solution' of the 'integral equation' $\theta(s) = \sigma s + b(s)$. This gives the result. □

The usefulness of the transformations in Theorem 3.1 may be illustrated by the following simple example.

Application 3.3. *The equation*

$$u_t = (u^m(1-u)^n u_x)_x + \begin{cases} u^{1-m}(1-u)^{1-n} & \text{for } 0 < u < 1 \\ 0 & \text{otherwise} \end{cases} \tag{3.11}$$

with $m > -1$ *and* $n > -1$ *admits a wavefront solution from* 1 *to* 0 *with wave speed* σ *if and only if* $\sigma \geq 2$.

Proof. From the pioneering work of Kolmogorov, Petrovskii and Piskunov [186], we know that the Fisher equation

$$u_t = u_{xx} + u(1-u) \tag{3.12}$$

admits a wavefront solution from 1 to 0 for any wave speed $\sigma \geq 2$ but no such solution for $\sigma < 2$. Noting that equation (3.11) has the form (3.1) with $a(u) = \int_0^u s^m(1-s)^n\, ds$ and $c(u) = u^{1-m}(1-u)^{1-n}$, and, equation (3.12) has the form (3.2) with $c(u)a'(u) = u(1-u)$, the result is immediate from the theorem. □

Note that in the converse case in Theorem 3.2 we are only able to say that equation (3.6) has a travelling-wave solution for *some* b satisfying (3.9) and not necessarily for any b satisfying (3.9). As an explicit illustration of where things can go wrong, choose a function $c(u) := u(\ell - u)(2u - \ell)$ for some ℓ. With $\sigma = 0$ the integral equation (3.10) admits the explicit solution $\theta(s) = s(\ell - s)$, and thus (3.8) has a stationary wavefront solution from ℓ to 0. At the same time, (3.9) is solved by $b = \pm\theta$. However of these functions, $b = \theta$ is the only one to which the conclusions of the theorem apply. The issue is further related to the question of whether or not (3.10) has a unique solution satisfying the integrability condition on $[0, \ell]$ with $\theta(\ell) = 0$; while, from the theory presented in Section 2.2, we know that in general no matter how smooth c is, this integral equation may have an infinite number of solutions.

Diverse variants on Theorems 3.1 and 3.2 are conceivable. For instance, it is possible to state similar results for semi-wavefront solutions and unbounded monotonic travelling-wave solutions. The above-stated results serve as examples of what is possible.

For completeness, we also mention the following.

Theorem 3.4. *Suppose that $\ell < \infty$. Equation* (1.1) *has a wavefront solution from ℓ to 0 with speed σ if and only if the equation*

$$u_t = (\widetilde{a}(u))_{xx} + (\widetilde{b}(u))_x + \widetilde{c}(u), \tag{3.13}$$

with

$$\widetilde{a}(u) = a(\ell) - a(\ell - u), \quad \widetilde{b}(u) = b(\ell - u) - b(\ell) \quad \text{and} \quad \widetilde{c}(u) = -c(\ell - u),$$

has a wavefront solution from ℓ to 0 with speed $-\sigma$. Moreover, in both instances the number of distinct wavefront solutions is the same.

This result can be obtained by direct examination of equation (1.1). It is easily checked that if u is a solution of (1.1), then $v(x,t) := \ell - u(-x,t)$ is a solution of (3.13), and, vice versa. Furthermore, if $u(x,t) \to \ell$ as $x \to -\infty$ and $u(x,t) \to 0$ as $x \to \infty$, then the same holds for v. Finally, if u has the form $u(x,t) = f(x - \sigma t)$ for some function f and constant σ, then v has the form $v(x,t) = g(x + \sigma t)$ where $g(\xi) = \ell - f(-\xi)$, and vice versa. In terms of the correspondence between wavefront solutions and solutions of the integral equation, Theorem 3.4 may also be viewed as a consequence of the next lemma and its proof.

Lemma 3.5. *Suppose that $\ell < \infty$. Then the following statements are equivalent: equation* (1.9) *has a solution θ on $[0,\ell]$ with $\theta(\ell) = 0$; the equation*

$$\Theta(s) = -\sigma s + \widetilde{b}(s) - \int_0^s \frac{\widetilde{c}(r)\widetilde{a}'(r)}{\Theta(r)}\, dr \tag{3.14}$$

has a solution Θ on $[0,\ell]$ in the sense of Definition 2.12 *with $\Theta(\ell) = 0$; and, equations* (1.9) *and* (3.14) *both have solutions on $[0,\ell]$. Idem ditto, the statements — equation* (1.9) *has a solution θ satisfying the integrability condition on $[0,\ell]$ with $\theta(\ell) = 0$; equation* (3.14) *has a solution Θ satisfying the integrability condition on $[0,\ell]$ with $\Theta(\ell) = 0$; and, equations* (1.9) *and* (3.14) *both have solutions satisfying the integrability condition on $[0,\ell]$ — are equivalent.*

Proof. The equivalence of the first two statements can be established quite simply by relating θ and Θ via $\theta(s) = \Theta(\ell - s)$. Whence, plainly each of the first two statements implies the third. To complete the proof of the lemma, suppose therefore that the third statement is true. Let θ and Θ denote the corresponding solutions of (1.9) and (3.14). Then since $\theta(0) = 0 \le \Theta(\ell)$ and $\theta(\ell) \ge 0 = \Theta(0)$, by continuity there must be a point $s^* \in [0,\ell]$ such that $\theta(s^*) = \Theta(\ell - s^*)$. Consequently the function

$$\psi(s) := \begin{cases} \theta(s) & \text{for } s \le s^* \\ \Theta(\ell - s) & \text{for } s > s^* \end{cases}$$

constitutes a solution of (1.9) on $[0,\ell]$ with $\psi(\ell) = 0$. Moreover if θ and Θ satisfy the integrability condition on $[0,\ell]$ so does ψ. Thus the first statement follows from the third. □

Bibliographical notes

The equivalence noted in Theorem 3.1 was derived independently under various restrictions by Hadeler [145], by Engler [97], and by Danilov et al. [78]. Furthermore, differentiating the relation $a(f_2(\xi)) = a(f_1(\Psi(\xi)))$ which follows from the transformation in Theorem 3.1, and using (3.3) to eliminate $\Psi'(\xi)$ yields $f_2'(\xi) = (a(f_1))'(\Psi(\xi))$ for all $-\infty < \xi < \infty$. This particular relation for the equivalence between travelling-wave solutions f_1 and f_2 of equations (3.1) and (3.2) was previously derived in a special case by Danilov et al. [78] for monotonic solutions defined in an arbitrary interval.

The transformations in Theorem 3.2 may be attributed to Engler [97]. Returning to the point that the converse case of Theorem 3.2 only states that equation (3.6) has a travelling-wave solution for *some* b satisfying (3.9) and not necessarily for any b satisfying (3.9); the inference in Engler's paper [97] is nonetheless that the latter is true. However careful reading shows that there too only the former statement can be justified.

Chapter 4

Travelling waves

In this chapter we report a number of general results on the existence of semi-wavefront solutions of equation (1.1) which can be obtained from the study of the integral equation (1.9).

4.1 Admissible wave speeds

Our first result states that the set of speeds for which such a solution decreasing to 0 exists is either empty, or, connected and unbounded above.

Theorem 4.1. *If equation* (1.1) *has a semi-wavefront solution with speed* σ_0 *decreasing to* 0, *then the equation has such a solution for every wave speed* $\sigma \geq \sigma_0$.

This theorem is an immediate consequence of a comparison principle for solutions of the integral equation, Lemma 2.18.

The above theorem may be extended to show how the existence of a semi-wavefront solution of one equation of the type (1.1) implies the existence of such a solution for another equation of the same type.

Theorem 4.2. *Consider equation* (1.1) *with two different sets of coefficients* a_i, b_i *and* c_i *on some interval* I *for* $i = 1, 2$. *Let* σ_1 *and* σ_2 *denote real parameters.*

(a) *Suppose that* $u \mapsto \sigma_2 u + b_2(u) - \sigma_1 u - b_1(u)$ *is a nondecreasing function on* $(0, \ell)$, *and,* $(c_2 a_2')(u) \leq (c_1 a_1')(u)$ *for all* $0 < u < \ell$.

(b) *Suppose that* $\sigma_2 u + b_2(u) \geq \sigma_1 u + b_1(u)$ *and* $\max\{0, (c_2 a_2')(u)\} \leq (c_1 a_1')(u)$ *for all* $0 < u < \ell$.

Then in both cases (a) *and* (b), *if equation* (1.1) *with* $i = 1$ *admits a semi-wavefront solution with speed* σ_1 *decreasing to* 0, *so does* (1.1) *with* $i = 2$ *and speed* σ_2.

Proof. Part (a) follows from Lemma 2.18(i). Part (b) follows in a similar pattern from Lemma 2.18(ii). □

It follows from Theorem 4.1 that there exists a $\sigma^*, -\infty \leq \sigma^* \leq \infty$, such that equation (1.1) has no semi-wavefront solution decreasing to 0 with any speed $\sigma < \sigma^*$ whereas it does have such a solution for every $\sigma > \sigma^*$. This critical wave speed may be characterized using a sort of 'variational principle', which is also a straightforward consequence of the theory of the integral equation (1.9).

Theorem 4.3. *Let $\mathcal{R}$ denote the set of nonnegative continuous functions ψ defined on an interval $[0,\iota)$ such that*

$$\int_{s_0}^{s_1} \frac{\{1+|c(r)|\}a'(r)}{\psi(r)}\,dr < \infty \quad \textit{for all } 0 < s_0 < s_1 < \iota$$

for some $0<\iota<\ell$, and let $\mathcal{S}$ denote the subset of $\psi \in \mathcal{R}$ such that $\psi(0)=0$ and

$$\lim_{\varepsilon \downarrow 0} \int_\varepsilon^s \frac{c(r)a'(r)}{\psi(r)}\,dr \quad \textit{exists and is finite for every } 0 < s < \iota.$$

Define the functionals

$$\mathcal{F}(\psi) = \sup_{0<r<s<\iota} \left\{ \frac{\psi(s)-\psi(r)-b(s)+b(r)+\int_r^s \frac{c(w)a'(w)}{\psi(w)}\,dw}{s-r} \right\}$$

and

$$\mathcal{G}(\psi) = \sup_{0<s<\iota} \left\{ \frac{\psi(s)-b(s)+\int_0^s \frac{c(r)a'(r)}{\psi(r)}\,dr}{s} \right\}$$

on $\mathcal{R}$. Set $\sigma^* := \inf\{\mathcal{F}(\psi) : \psi \in \mathcal{S}\}$. Then equation (1.1) *has a semi-wavefront solution decreasing to* 0 *for all speeds $\sigma > \sigma^*$, such a solution with speed σ^* if and only if $\sigma^* = \mathcal{F}(\psi)$ for some $\psi \in \mathcal{S}$, and, no such solution for any speed $\sigma < \sigma^*$. Furthermore when $c(u) \geq 0$ for all $0 < u < \delta$ for some $0 < \delta \leq \ell$ there holds $\sigma^* = \inf\{\mathcal{G}(\psi) : \psi \in \mathcal{R}\}$, and, $\sigma^* = \mathcal{F}(\psi)$ for some $\psi \in \mathcal{S}$ if and only if $\sigma^* = \mathcal{G}(\psi)$ for some $\psi \in \mathcal{R}$.*

Proof. Suppose to begin with that equation (1.1) does have a solution of the desired type with speed σ. Then by Theorem 2.9 the integral equation (1.9) has a solution θ satisfying the integrability condition on an interval $[0,\delta)$ for some $0<\delta<\ell$. Subsequently, by definition $\theta \in \mathcal{S}$, and by (1.9) there holds

$$\sigma(s-r) = \theta(s)-\theta(r)-b(s)+b(r)+\int_r^s \frac{c(w)a'(w)}{\theta(w)}\,dw$$

for all $0<r<s<\delta$, i.e. $\mathcal{F}(\theta)=\sigma$. Next, let σ be such that there is a $\psi \in \mathcal{S}$ with $\sigma \geq \mathcal{F}(\psi)$. Let $[0,\iota)$ denote the domain of definition of ψ and

$$\widetilde{b}(s) := \psi(s) - \sigma s + \int_0^s \frac{c(r)a'(r)}{\psi(r)}\,dr.$$

This means that ψ is a solution of equation (1.9) with b replaced by $\widetilde{b}$ on $[0,\iota)$, while the function $b-\widetilde{b}$ is nondecreasing on $[0,\iota)$. Subsequently, by Lemma 2.18(i), the integral equation (1.9) has a solution θ on $[0,\iota)$ such that $\theta(s) \geq \psi(s)$ for all $0<s<\iota$. Whence θ also satisfies the integrability condition on $[0,\iota)$, and by Theorem 2.9 equation (1.1) has a travelling-wave solution of the desired type with wave speed σ. This proves the main conclusion of the theorem. The conclusion when $c \geq 0$ on $(0,\delta)$ for some $0<\delta\leq\ell$ may be obtained by repeating the above argument with Lemma 2.18(ii) in lieu of Lemma 2.18(i). $\square$

It is possible that equation (1.1) admits a semi-wavefront solution decreasing to 0 for all wave speeds ($\sigma^* = -\infty$), some wave speeds ($-\infty < \sigma^* < \infty$), and no wave speeds ($\sigma^* = \infty$). Furthermore, when the critical value σ^* is finite, equation (1.1) may or may not admit such a solution with this specific speed. Illustrations of all of these possibilities will be provided in the coming chapters.

4.2 Number of solutions

Concerning the number of semi-wavefront solutions of a general equation of the class (1.1) with any particular wave speed, we can state the following.

Theorem 4.4. *For any fixed wave speed σ, equation* (1.1) *has at most one distinct semi-wavefront solution decreasing to* 0 *whenever any one of the following hold.*

(a) $c(u) < 0$ *for all* $0 < u < \ell$.

(b) ca' *is differentiable in* $(0, \ell)$ *and* $c(u) \leq 0$ *for all* $0 < u < \delta$ *for some* $0 < \delta < \ell$.

Definition 4.5. The partial differential equation (1.1) will be said to admit a one-parameter family of distinct semi-wavefront solutions with wave speed σ decreasing to 0 when there exists a continuous order-preserving bijective mapping from the interval $(0, 1]$ onto the set of all such solutions.

Theorem 4.6. *For any fixed wave speed σ, equation* (1.1) *has either a one-parameter family of distinct semi-wavefront solutions decreasing to* 0 *in the sense of Definition* 4.5 *or no such solution whenever any one of the following hold.*

(a) $c(u) > 0$ *for all* $0 < u < \ell$.

(b) ca' *is differentiable in* $(0, \ell)$ *and* $c(u) > 0$ *for all* $0 < u < \delta$ *for some* $0 < \delta < \ell$.

In the coming chapters we shall see that in accordance with Theorems 4.4 and 4.6 all three alternatives implied by these theorems are indeed possible, i.e. equation (1.1) may have no semi-wavefront solution with speed σ decreasing to 0, exactly one distinct solution of this type, or, a one-parameter family of such solutions.

To prove Theorems 4.4 and 4.6, we use the next two lemmata.

Lemma 4.7. *Consider equation* (1.9) *with two different wave speeds σ_i, sets of coefficients a_i, b_i and c_i, and corresponding solutions θ_i on an interval $[0, \delta]$ with $\delta \leq \ell$ and $\delta < \infty$ for $i = 1, 2$. Suppose that the function $s \mapsto \sigma_2 s + b_2(s) - \sigma_1 s - b_1(s)$ is nondecreasing on $[0, \delta]$, $c_2 a_2' \leq c_1 a_1'$ almost everywhere in $(0, \delta)$, and, $\max\{\theta_1(s), \theta_2(s)\} > 0$ for all $0 < s < \delta$. Then if there exists a point $s^* \in (0, \delta)$ such that $\theta_2(s^*) \geq \theta_1(s^*)$, there holds $\theta_2 \geq \theta_1$ on $[s^*, \delta]$. Moreover, if $c_1 \geq 0$ on (s^*, δ), then either $\theta_2(\delta) > \theta_1(\delta)$ or $\theta_2 \equiv \theta_1$ on $[s^*, \delta]$.*

Proof. Equation (1.9) can be rewritten as

$$\theta(s) = \theta(s^*) + \sigma_i s + b_i(s) - \sigma_i s^* - b_i(s^*) - \int_{s^*}^{s} \frac{c_i(r) a_i'(r)}{\theta(r)} \, dr \tag{4.1}$$

for $s \in [s^*, \delta)$ and $i = 1, 2$. Subsequently, applying Lemma 2.18(i) to this reformulation, from the existence of θ_1 and θ_2 we deduce the existence of a solution θ^* of (4.1) for $i = 2$ on $[s^*, \delta)$ such that $\theta^*(s) \geq \max\{\theta_1(s), \theta_2(s)\}$ for all $s \in [s^*, \delta)$. However, since by Lemma 2.13 positive solutions of (4.1) are unique, necessarily $\theta^* \equiv \theta_2$. This proves the first assertion of the lemma. The remaining assertion subsequently follows from Lemma 2.18(ii). □

Lemma 4.8. *Suppose that* $c(u) > 0$ *for all* $0 < u < \delta$ *for some* $0 < \delta < \ell$ *and that equation* (1.9) *admits at least one solution on* $[0, \delta)$. *Then, for every* $0 < \varrho \leq 1$ *equation* (1.9) *admits a unique solution* $\theta(\cdot; \varrho)$ *on an interval* $[0, \Delta(\varrho)]$ *such that the mapping* $\varrho \mapsto (\theta(\Delta(\varrho); \varrho), \Delta(\varrho))$ *is continuous, given any* $0 < \varrho_1 < \varrho_2 \leq 1$ *there holds* $0 < \Delta(\varrho_1) \leq \Delta(\varrho_2) \leq \Delta(1) = \delta$ *and* $\theta(s; \varrho_1) < \theta(s; \varrho_2)$ *for all* $0 < s \leq \Delta(\varrho_1)$, *and, no other solutions.*

Proof. Since equation (1.9) admits a solution on $[0, \delta)$, it has a maximal solution θ^* on $[0, \delta]$ by Lemmata 2.14 and 2.17. We set $\Gamma := \{(s, 0) : 0 < s \leq \delta\} \cup \{(\delta, \rho) : 0 < \rho \leq \theta^*(\delta)\}$ and let $\varrho \mapsto (\Delta(\varrho), \vartheta(\varrho))$ be a homeomorphism $(0, 1] \to \Gamma$ with the property that $\Delta(\varrho_1) \leq \Delta(\varrho_2)$ and $\vartheta(\varrho_1) \leq \vartheta(\varrho_2)$ for every $0 < \varrho_1 < \varrho_2 \leq 1$. By Lemma 2.18 any solution θ of equation (1.9) is defined on a maximal subinterval of $[0, \delta]$, $[0, \delta^*]$ say, with $(\delta^*, \theta(\delta^*)) \in \Gamma$. On the other hand, by Lemma A3(i) the equation

$$\theta(s) = \vartheta(\varrho) + \sigma s + b(s) - \sigma\Delta(\varrho) - b(\Delta(\varrho)) + \int_s^{\Delta(\varrho)} \frac{c(r)a'(r)}{\theta(r)}\, dr$$

has a unique solution $\theta(\cdot; \varrho)$ on $[0, \Delta(\varrho)]$ which is positive on $(0, \Delta(\varrho))$ for every $0 < \varrho \leq 1$. Moreover, by Lemma 4.7, given any $0 < \varrho_1 < \varrho_2 \leq 1$ there holds $\theta(s; \varrho_1) < \theta(s; \varrho_2)$ for all $0 < s \leq \Delta(\varrho_1)$. □

Proof of Theorems 4.4 *and* 4.6. If $c(s) \leq 0$ for all $0 < s < \delta$ for some $0 < \delta < \ell$, then equation (1.9) has at most one solution on $[0, \delta)$, by Lemma 2.14(i). On the other hand, if $c(s) > 0$ for all $0 < s < \delta$ for some $0 < \delta < \ell$, by Lemma 4.8, either equation (1.9) has no solution, or, there is a continuous order-preserving bijective mapping from $(0, 1]$ onto the set of solutions. Moreover, in the latter case, every solution is positive in the interior of its maximal interval of existence contained in $[0, \delta)$. Lemmata 2.25, 2.40 and 2.24 subsequently give the desired results. □

Bibliographical notes

Additional details of the proof of Theorem 4.2 can be found in [124]. Theorem 4.3 is a generalization of the 'variational principle' for wavefront solutions of (1.1) discovered by Hadeler and Rothe [150] and also described in [141–145, 295, 297].

Chapter 5

Convection-diffusion

Considering only diffusion-convection processes, i.e. the equation

$$u_t = (a(u))_{xx} + (b(u))_x, \tag{5.1}$$

the integral equation (1.9) reduces to the simple identity $\theta(s) = \sigma s + b(s)$. Moreover, by Lemma 2.40 any such 'solution' θ satisfies the integrability condition in an interval $[0, \delta)$ if and only if it is positive in $(0, \delta)$. The search for nonnegative solutions of the integral equation satisfying the integrability condition is therefore reduced to the search for σ such that $\sigma s + b(s) > 0$ for all $0 < s < \delta$ for some $0 < \delta \leq \ell$. This leads readily to the next result.

Theorem 5.1 (Existence). *Let*

$$\sigma^* := \limsup_{s \downarrow 0} \frac{-b(s)}{s}.$$

Then for every wave speed σ equation (5.1) has either exactly one distinct semi-wavefront solution decreasing to 0 or no such solution. Moreover, the equation has such a solution for all $\sigma > \sigma^*$, such a solution for $\sigma = \sigma^*$ if and only if

$$\sigma^* s + b(s) > 0 \quad \textit{for all } 0 < s < \delta \tag{5.2}$$

for some $0 < \delta \leq \ell$, and no such solution for all $\sigma < \sigma^*$.

With regard to the boundedness of the support of the travelling-wave solutions the following can be stated.

Theorem 5.2 (Bounded support). *Suppose that the conditions of Theorem 5.1 hold.*

(i) *If*

$$\int_0^\delta \frac{a'(s)}{\max\{s, b(s)\}}\, ds = \infty \quad \textit{for some } 0 < \delta < \ell, \tag{5.3}$$

then every semi-wavefront solution decreasing to 0 is positive everywhere in its domain of definition.

(ii) *If*

$$\int_0^\delta \frac{a'(s)}{\max\{s, b(s)\}}\, ds < \infty \quad \textit{for some } 0 < \delta < \ell, \tag{5.4}$$

then every solution of this type with wave speed $\sigma > \sigma^*$ is such that its support is bounded above.

Proof. For any positive function $\theta(s) = \sigma s + b(s)$ on an interval $(0, \delta)$, we can estimate $\theta(s) \leq (|\sigma| + 1)\max\{s, b(s)\}$ for $0 < s < \delta$. Whence if (5.3) holds, so does (2.37) and Corollary 2.30 yields part (i) of the theorem. On the other hand, if $\sigma > \sigma^*$, then choosing $\sigma > \sigma_0 > \sigma^*$, $\gamma := \max\{\sigma - \sigma_0, 1, -\sigma_0\}$ and $\delta_0 > 0$ so small that $\sigma_0 s + b(s) > 0$ for all $0 < s < \delta_0$, we can estimate

$$\begin{aligned}
\theta(s) &= (\sigma - \sigma_0)s + \sigma_0 s + b(s) \\
&= (\sigma - \sigma_0)s + \max\{0, \sigma_0 s + b(s)\} \\
&\geq (\sigma - \sigma_0)s + \frac{\sigma - \sigma_0}{\gamma}\max\{0, \sigma_0 s + b(s)\} \\
&= \frac{\sigma - \sigma_0}{\gamma}\max\{\gamma s, (\gamma + \sigma_0)s + b(s)\} \\
&\geq \frac{\sigma - \sigma_0}{\gamma}\max\{s, b(s)\}
\end{aligned}$$

for $0 < s < \delta_0$. Whence if $\sigma > \sigma^*$ and (5.4) holds, so does (2.35) for small enough δ, and Corollary 2.30 yields part (ii) of the theorem. □

In the event that (5.4) holds and the convection-diffusion equation (5.1) has a semi-wavefront solution with the critical wave speed σ^* decreasing to 0, this semi-wavefront may or may not have bounded support. In general, when (5.2) and (5.4) hold, for the critical wave speed σ^* there is no other option but to test the criterion (2.35) explicitly, i.e.

$$\int_0^\delta \frac{a'(s)}{\sigma^* s + b(s)}\, ds < \infty \qquad \text{for some } 0 < \delta \leq \ell.$$

As an application of the above theorems, let us examine the model for the one-dimensional flow of a thin viscous film over a flat plate introduced in [57]. After a suitable normalization this model takes the form of the equation

$$u_t = (u^4)_{xx} + b_0(u^3)_x, \tag{5.5}$$

in which t denotes time, u denotes the thickness of the fluid film, and b_0 is the angle of inclination of the plate in the direction x. The mechanism behind this model is the force of gravity which acts in two ways. The first is that it tends to drive the fluid from regions of greater thickness to those where the film is thinner. This accounts for the diffusive second-order term on the right-hand side of the equation. The second is that it tends to pull the fluid to a lower elevation. This accounts for the first-order term on the right-hand side of the equation. If the plate is horizontal $b_0 = 0$, if the plate is inclined uphill $b_0 > 0$, and, if the plate is inclined downhill $b_0 < 0$. Of particular interest in this model is the existence of waves with bounded support, for, recalling that implicitly where a fluid film is present $u > 0$ and where there is no fluid present $u = 0$, the boundary of the support of a solution demarcates a leading edge of the flow.

Application 5.3. *For every wave speed σ equation* (5.5) *has at most one distinct semi-wavefront solution decreasing to* 0, *and the support of this solution is necessarily bounded above.*

(i) *If $b_0 > 0$, then the equation admits such a solution if and only if $\sigma \geq 0$.*

(ii) *If $b_0 \leq 0$, then the equation admits such a solution if and only if $\sigma > 0$.*

Thus all semi-wavefront solutions representing a fluid film which may approach a zero thickness necessarily model a phenomenon in which there is a free boundary between a domain where the fluid film is present and one where it is absent. Of note too is the interpretation which may be given to the admissible range of wave speeds. The first point is that there are no waves with negative speed. This would infer that the diffusive mechanism is strong enough to ensure that the surface of the plate covered by the fluid film cannot contract. The second point is the possible occurrence of what is essentially a stationary solution with wave speed $\sigma = 0$. If $b_0 > 0$, then such a solution occurs, but if $b_0 \leq 0$ it does not. This would imply that if the plate is inclined uphill, the downward gravitational pull may be strong enough to balance the diffusive mechanism. On the other hand, if the plate is horizontal or inclined downhill, then the motion will always tend to enlarge the surface covered by the fluid.

As a contrast to the properties of equation (5.5), one might like to consider the Burgers equation in a similar guise,

$$u_t = u_{xx} + b_0(u^2)_x. \tag{5.6}$$

Application 5.4. *For every wave speed σ equation* (5.6) *has at most one distinct semi-wavefront solution decreasing to* 0, *and this solution is necessarily positive everywhere in its domain of definition.*

(i) *If $b_0 > 0$, then the equation admits such a solution if and only if $\sigma \geq 0$.*

(ii) *If $b_0 \leq 0$, then the equation admits such a solution if and only if $\sigma > 0$.*

Noting that in the Burgers equation the unknown u denotes a variable which is a measure of turbulence in hydrodynamic flow, the inference of the positivity of the semi-wavefront solutions is that in it is impossible to have regions with turbulence in combination with regions where there is no dissipation of energy.

A further illustration is provided by the foam drainage equation,

$$u_t = (u^{3/2})_{xx} + b_0(u^2)_x. \tag{5.7}$$

Modelling the motion of a foam composed of gas bubbles trapped in a liquid in one spatial dimension, the unknown in this equation denotes the liquid fraction of the foam, t time, and, x a spatial coordinate. Analogous to the viscous-film model, the parameter b_0 is positive if x decreases in the direction of gravitational pull, negative if it increases in the direction of gravitational pull, and, zero if the motion occurs in a horizontal direction.

Application 5.5. *For every wave speed σ equation* (5.7) *has at most one distinct semi-wavefront solution decreasing to* 0.

(i) *If $b_0 > 0$, then the equation admits such a solution if and only if $\sigma \geq 0$. Moreover, if $\sigma = 0$ this solution is necessarily positive everywhere in its domain of definition. Whereas, if $\sigma > 0$ the support of this solution is necessarily bounded above.*

(ii) *If $b_0 \leq 0$, then the equation admits a semi-wavefront solution decreasing to* 0 *if and only if $\sigma > 0$, and the support of this solution is necessarily bounded above.*

This example invites a similar interpretation to Application 5.3. However, the fact that in the present example in the case $b_0 > 0$ the semi-wavefront solution with wave speed $\sigma = 0$ does not have bounded support indicates that for the foam drainage equation the diffusive mechanism is a little stronger in comparison to the gravitational pull than for the viscous-film model.

We leave it to the reader to verify the above three examples. Note that they all illustrate that in Theorem 5.1, equation (5.1) may or may not have a semi-wavefront solution with the critical wave speed σ^* decreasing to 0. In each example $\sigma^* = 0$, but it is the sign of b_0 which determines whether or not (5.2) holds for some $0 < \delta \leq \infty$ and thus whether or not there is such a solution. Parts (i) of Applications 5.3 and 5.5 also illustrate that, when the convection-diffusion equation (5.1) has a semi-wavefront solution with the critical wave speed σ^* and the conclusions of Theorem 5.2(ii) hold, the distinct semi-wavefront with the critical speed may or may not have bounded support.

Bibliographical notes

Theorem 5.1 may be found in [119], while Theorem 5.2 is based on the refinement of estimates in [119, 127].

The derivation of the model (5.5) for the flow of a thin viscous film considered is due to Buckmaster and is described in full in [57]. For further information on the derivation of the Burgers equation (5.6) and properties of solutions of this equation, we refer to the original publications of Burgers [60–63]. Correspondingly, further information on the origin of the foam drainage equation (5.7) can be found in [131, 292, 293, 304]. Similar and more general equations of the form (5.1) also arise in the theory of soil-moisture flow [30, 68, 122, 130, 238, 278]. In this context, equation (5.5) is often called the *Richards equation* [122, 246] or the *nonlinear Fokker–Planck equation* [30, 238].

Each of the Applications 5.3 – 5.5 is followed by a discussion of certain inferences for the underlying nonlinear convection-diffusion equation. These are drawn from the conclusion of whether or not the support of the travelling-wave solution is bounded above. In fact these observations have been mathematically rigorously proven to hold for arbitrary solutions of the respective equations using the semi-wavefront solutions and a comparison principle [116, 119, 121].

Chapter 6

Reaction-diffusion

The class of equations of the type (1.1) in which the convection term is absent, i.e. for which the equation has the form

$$u_t = (a(u))_{xx} + c(u), \tag{6.1}$$

encompasses the Fisher equation, the Newell–Whitehead equation, the Zeldovich equation, the KPP equation, the Nagumo equation, and many other commonly-used models of diffusion-reaction processes. In these models the reaction term does not change sign in a right neighbourhood of zero, and, generally the coefficients a and c are smooth. For an equation of the class (6.1) the corresponding integral equation (1.9) reduces to

$$\theta(s) = \sigma s - \int_0^s \frac{c(r)a'(r)}{\theta(r)}\, dr. \tag{6.2}$$

Moreover, when c has a fixed sign near zero or when ca' is sufficiently smooth, this equation possesses a structure which is relatively convenient for analysis. In this chapter, we shall utilize this structure to examine the existence of semi-wavefront solutions of the reaction-diffusion equation (6.1) when c has a definite sign near zero, and, when ca' is continuously differentiable in a right neighbourhood of zero. Furthermore, we shall identify circumstances under which such a solution is positive everywhere or may have bounded support.

6.1 Sink term

We begin with the case that the reaction term in (6.1) is a definite absorption or sink term.

Theorem 6.1 (Existence). *Suppose that $c(u) < 0$ for all $0 < u < \ell$. Then for every wave speed σ equation* (6.1) *has exactly one distinct semi-wavefront solution decreasing to* 0.

Proof. By Lemma 2.16(i) of the theory of the integral equation, (6.2) has a unique solution on $[0, \ell)$ for every σ. Moreover, this solution is positive on $(0, \ell)$. The present theorem is then an immediate consequence of Theorems 2.9 and 4.4. □

According to Corollary 2.30, whether or not the semi-wavefront solution in Theorem 6.1 has bounded support depends on the behaviour of the solution θ

of the integral equation (6.2) as $s \downarrow 0$. As we shall see, this behaviour is in turn largely determined by that of the variable

$$Q(s) := \left| 2 \int_0^s c(r) a'(r)\, dr \right|^{1/2}. \tag{6.3}$$

To be specific, invoking Corollary 2.30 we shall prove the following.

Theorem 6.2 (Bounded support). *Suppose that the conditions of Theorem 6.1 hold. Fix $0 < \delta < \ell$.*

(a) *If*

$$\int_0^\delta \frac{a'(s)}{\max\{Q(s), s\}}\, ds = \infty,$$

then every semi-wavefront decreasing to 0 is positive everywhere in its domain of definition.

(b) *If*

$$\int_0^\delta \frac{a'(s)}{\max\{Q(s), s\}}\, ds < \infty$$

and

$$\int_0^\delta \frac{a'(s)}{Q(s)}\, ds = \infty, \tag{6.4}$$

then every solution of this type with wave speed $\sigma \le 0$ is positive everywhere in its domain of definition, and every solution of this type with wave speed $\sigma > 0$ is such that its support is bounded above.

(c) *If*

$$\int_0^\delta \frac{a'(s)}{Q(s)}\, ds < \infty, \tag{6.5}$$

then every solution of this type with wave speed $\sigma \ge 0$ is such that its support is bounded above. Moreover, if ca' is absolutely continuous in $[0, \delta)$, $(ca')(0) = 0$, $\lim \operatorname{ess\,sup}_{s\downarrow 0} (ca')'(s) \le 0$, and,

$$\int_0^\delta \frac{1}{|c(s)|}\, ds = \infty,$$

then every solution of this type with wave speed $\sigma < 0$ is positive everywhere in its domain of definition. On the other hand, if $\limsup_{s\downarrow 0} Q(s)/s = \liminf_{s\downarrow 0} Q(s)/s > 0$, if $\lim \operatorname{ess\,inf}_{s\downarrow 0} c(s)a'(s)/Q(s) > 0$, or, if ca' is absolutely continuous in $[0, \delta)$, $(ca')(0) = 0$, $\lim \operatorname{ess\,inf}_{s\downarrow 0} (ca')'(s) > -\infty$, and,

$$\int_0^\delta \frac{1}{|c(s)|}\, ds < \infty,$$

then every solution of this type with wave speed $\sigma < 0$ is such that its support is bounded above.

Recalling Corollary 2.30, this theorem follows directly from the next lemma.

Lemma 6.3. *Suppose that $c(u) \le 0$ and $Q(u) > 0$ for all $0 < u < \delta$ for some $0 < \delta < \ell$. Suppose furthermore that equation (6.2) has a unique solution θ on $[0, \delta)$.*

(i) *If $\sigma = 0$, then $\theta(s) = Q(s)$ for all $0 < s < \delta$.*

(ii) *If $\sigma > 0$, then*

$$\min\{\sigma, 1\} \max\{s, Q(s)\} \le \theta(s) \le (\sigma + 1) \max\{s, Q(s)\} \tag{6.6}$$

for all $0 < s < \delta$.

(iii) *If $\sigma \neq 0$, and, $Q(s)/s \to \mu$ as $s \downarrow 0$ for some $0 \le \mu \le \infty$, then*

$$\frac{\theta(s)}{s} \to \frac{\sigma + \sqrt{\sigma^2 + 4\mu^2}}{2} \qquad \text{as } s \downarrow 0. \tag{6.7}$$

Moreover, if $\sigma > 0$ or $\mu > 0$, then

$$\frac{\theta(s)}{Q(s)} \to \frac{\sigma/\mu + \sqrt{\sigma^2/\mu^2 + 4}}{2} \qquad \text{as } s \downarrow 0. \tag{6.8}$$

(iv) *If $\sigma < 0$, and, $|c(s)a'(s)| \ge AQ(s)$ for almost all $0 < s < \delta$ for some $A > 0$, then*

$$\theta(s) \ge \frac{\sigma + \sqrt{\sigma^2 + 4A^2}}{2A} Q(s) \qquad \text{for all } 0 < s < \delta.$$

(v) *If $\sigma < 0$, ca' is absolutely continuous on $[0, \delta)$, $(ca')(0) = 0$, and, $(ca')'(s) \ge A$ for almost all $0 < s < \delta$ for some $A \le 0$, then*

$$\theta(s) \ge \frac{2}{\sqrt{\sigma^2 - 4A} - \sigma} |c(s)a'(s)| \qquad \text{for all } 0 < s < \delta.$$

(vi) *If $\sigma < 0$, ca' is absolutely continuous on $[0, \delta)$, $(ca')(0) = 0$, and, $(ca')'(s) \le B$ for almost all $0 < s < \delta$ for some $B \le \sigma^2/4$, then*

$$\theta(s) \le \frac{2}{\sqrt{\sigma^2 - 4B} - \sigma} |c(s)a'(s)| \qquad \text{for all } 0 < s < \delta.$$

Proof. We rely on the theory of the integral equation presented in Section 2.2.

(i) Taking $\alpha = \beta = 0$ in Lemma 2.16(ii) yields the desired result.

(ii) Taking $\alpha = 0$ and $\beta = \infty$ in Lemma 2.16(ii) yields

$$\sigma s \le \theta(s) \le \sigma s + Q(s) \qquad \text{for all } 0 \le s < \delta. \tag{6.9}$$

However, by Lemma 2.18(i), Lemma 2.14(i) and part (i) of the present lemma we also know

$$\theta(s) \ge Q(s) \qquad \text{for all } 0 \le s < \delta. \tag{6.10}$$

Combining (6.9) and (6.10) yields (6.6).

(iii) Pick $0 < \delta^* < \delta$. Define $A := \inf\{Q(s)/s : 0 < s < \delta^*\}$ and $B := \sup\{Q(s)/s : 0 < s < \delta^*\}$. Then, if $\sigma > 0$, taking $\alpha := \sigma/B$ and $\beta := \sigma/A$ in Lemma 2.16(ii) gives

$$\theta(s) \geq \sigma s + \frac{-\sigma/A + \sqrt{\sigma^2/A^2 + 4}}{2} Q(s) \tag{6.11}$$

and

$$\theta(s) \leq \sigma s + \frac{-\sigma/B + \sqrt{\sigma^2/B^2 + 4}}{2} Q(s) \tag{6.12}$$

for every $0 < s < \delta^*$. On the other hand, if $\sigma < 0$ and $\mu > 0$, choosing δ^* so small that $-A < \sigma\sqrt{1 - A/B}$, and, taking $\alpha = \sigma/A$ and $\beta = \sigma/B$ in Lemma 2.16(ii), we obtain, for any $0 < s < \delta^*$, (6.11) and (6.12) with the inequalities reversed. The assertions (6.7) and (6.8) subsequently follow in either case by passing to the limit $\delta^* \downarrow 0$. It therefore only remains to establish (6.7) when $\sigma < 0$ and $\mu = 0$. However, in this case, by Lemma 2.14(i), Lemma 2.18(i) and part (i) of the present lemma, there holds $\theta(s) \leq Q(s)$ for all $0 < s < \delta$. Whence the result is immediate.

(iv) It can be verified that $\theta_1(s) := \{(\sigma + \sqrt{\sigma^2 + 4A^2})/2A\}Q(s)$ is a solution of (1.9) on $[0, \delta)$ with $\sigma s + b(s)$ replaced by $\sigma_1 s + b_1(s) := (\sigma/A)Q(s)$. Moreover, $s \mapsto \sigma s - \sigma_1 s - b_1(s)$ is nondecreasing on $[0, \delta)$. Whence, by Lemma 2.18(i) and Lemma 2.14(i) there holds $\theta \geq \theta_1$ on $[0, \delta)$.

(v) The proof of this part is similar to that of part (iv), noting that $\theta_1(s) := \{2/(\sqrt{\sigma^2 - 4A} - \sigma)\}\,|c(s)a'(s)|$ is a solution of (1.9) on $[0, \delta)$ with $\sigma s + b(s)$ replaced by $\sigma_1 s + b_1(s) := \theta_1(s) + \{(\sigma - \sqrt{\sigma^2 - 4A})/2\}s$.

(vi) The function $\theta_2(s) := \{2/(\sqrt{\sigma^2 - 4B} - \sigma)\}\,|c(s)a'(s)|$ is the unique solution of (1.9) on $[0, \delta)$ with $\sigma s + b(s)$ replaced by $\sigma_2 s + b_2(s) := \theta_2(s) + \{(\sigma - \sqrt{\sigma^2 - 4B})/2\}s$. Moreover, $s \mapsto \sigma_2 s + b_2(s) - \sigma s$ is nondecreasing on $[0, \delta)$. So, by Lemma 2.18(i), necessarily $\theta \leq \theta_2$ on $[0, \delta)$. □

The equation

$$u_t = u_{xx} - H(u), \tag{6.13}$$

where H denotes the Heaviside function

$$H(u) = \begin{cases} 0 & \text{for } u \leq 0 \\ 1 & \text{for } u > 0, \end{cases}$$

has been derived as a model for the diffusion of oxygen in absorbing tissue. In this model, t denotes time, x distance into the tissue from its outer surface, and u the concentration of the oxygen which is free to diffuse. The sink term in the equation simulates the uptake of this oxygen by the tissue. The boundary of the support of any solution will denote the limit of the penetration of the free oxygen. The above theorems have the following significance for this equation.

Application 6.4. *For every wave speed σ equation (6.13) admits precisely one distinct semi-wavefront solution decreasing to 0. Moreover, the support of this solution is necessarily bounded above.*

Proof. For equation (6.13) we discern that $a'(s) = 1$ while the variable (6.3) is given by $Q(s) = \sqrt{2s}$ for all $s > 0$. Subsequently, (6.5) holds, and, $Q(s)/s \to \infty$ as $s \downarrow 0$. The assertion is now a corollary of Theorem 6.1 and Theorem 6.2 part (iii). □

The existence of a semi-wavefront solution with bounded support with every wave speed σ signifies that, without a sufficient supply of free oxygen at the surface of the tissue, the depth of penetration of unabsorbed oxygen will recede towards the surface.

Worthy of note is also that equation (6.13) may be derived as a reformulation of the classical Stefan problem. In this setting, the boundary of the support of the semi-wavefront solution corresponds to the usual free boundary in the Stefan problem.

6.2 Source term

When the reaction term in equation (6.1) is a source term, the situation is more complex than that described in the previous section. For a start, the equation need not always admit a semi-wavefront solution. This is determined by the finiteness of the parameter

$$\lambda_1 := \limsup_{s\downarrow 0} \left\{ \frac{1}{s} \int_0^s \frac{c(r)a'(r)}{r} \, dr \right\}. \tag{6.14}$$

In general for equation (6.1) with a source term there may hold $0 \leq \lambda_1 \leq \infty$. This may be verified by considering

$$u_t = u_{xx} + u^p \quad \text{where } p > 0.$$

If $p > 1$ then $\lambda_1 = 0$, if $p = 1$ then $\lambda_1 = 1$, and if $p < 1$ then $\lambda_1 = \infty$. Also, in general λ_1 need not be equal to

$$\lambda_0 := \liminf_{s\downarrow 0} \left\{ \frac{1}{s} \int_0^s \frac{c(r)a'(r)}{r} \, dr \right\}. \tag{6.15}$$

This can be checked by considering

$$u_t = u_{xx} + u(2 + \sin|\ln u| - \cos|\ln u|).$$

For this equation $\lambda_1 = 3$ whereas $\lambda_0 = 1$. Another complication, when the reaction term in equation (6.1) is a source term, is that the equation may admit more than one semi-wavefront solution with any given wave speed.

The following is the case, where

$$\Phi(z_0, z_1) := \begin{cases} (2z_1 - z_0)/\sqrt{2(z_1 - z_0)} & \text{for } 0 \le 3z_0 < 2z_1 \\ 2\sqrt{z_0} & \text{for } 0 \le 2z_1 \le 3z_0 \end{cases}$$

and

$$\Lambda_1 := \sup_{0<s<\delta} \left\{ \frac{1}{s} \int_0^s \frac{c(r)a'(r)}{r}\, dr \right\}. \tag{6.16}$$

Theorem 6.5 (Existence). *Suppose that $c(u) > 0$ for all $0 < u < \ell$. Then for every wave speed σ equation* (6.1) *has a one-parameter family of distinct semi-wavefront solutions decreasing to* 0 *in the sense of Definition* 4.5 *or no such solution. Moreover:*

(i) *When $\lambda_1 = \infty$ the equation has no such solution for all σ.*

(ii) *When $0 < \lambda_1 < \infty$ there exists a $\sigma^* > 0$ such that the equation has a one-parameter family of such solutions for all $\sigma > \sigma^*$ and no such solution for all $\sigma < \sigma^*$. The critical wave speed satisfies the inequalities $\Phi(\lambda_0, \lambda_1) \le \sigma^* \le 2\sqrt{\lambda_1}$. Furthermore, if $\lambda_0 = \Lambda_1$ for some $0 < \delta < \ell$, the equation has a one-parameter family of distinct semi-wavefront solutions decreasing to* 0 *with the critical speed $\sigma^* = 2\sqrt{\lambda_1}$.*

(iii) *When $\lambda_1 = 0$ the equation has a one-parameter family of such solutions for all $\sigma > 0$ and no such solution for all $\sigma \le 0$.*

Note that, in general when $0 < \lambda_1 < \infty$, equation (6.1) may or may not admit semi-wavefront solutions decreasing to 0 for the critical wave speed σ^*. By way of illustration, consider the equation

$$u_t = u_{xx} + u(1 + k\,|\ln u|^{-2}) \tag{6.17}$$

with $0 < \ell < 1$ and k a real parameter. For this equation it can be computed that $\lambda_0 = \lambda_1 = 1$ and thus that the critical wave speed is $\sigma^* = 2$ irrespective of the value of k. Moreover, $\Lambda_1 = 1$ for each $0 < \delta < \ell$ if and only if $k \le 0$. However, by Lemma 2.15 the equation admits a semi-wavefront solution with wave speed σ^* decreasing to 0 if and only if $k \le 1/4$.

The first assertion of Theorem 6.5 follows from Theorem 4.6. The remaining assertions are justified by the lemma below, when one bears in mind that without any loss of generality one can choose δ arbitrarily small.

Lemma 6.6. *Suppose that $c(u) > 0$ for all $0 < u < \delta$ for some $0 < \delta < \ell$. Define Λ_1 by* (6.16) *and*

$$\Lambda_0 := \inf_{0<s<\delta} \left\{ \frac{1}{s} \int_0^s \frac{c(r)a'(r)}{r}\, dr \right\}. \tag{6.18}$$

(i) *If $\sigma \le 0$ or $\sigma < 2\sqrt{\Lambda_0}$, equation* (6.2) *has no solution on $[0, \delta)$. Furthermore, given any solution θ of* (6.2) *on $[0, \delta)$ there holds $\theta(s) \le \mu s$ for all $0 < s < \delta$, where*

$$\mu := \frac{\sigma + \sqrt{\sigma^2 - 4\Lambda_0}}{2}.$$

(ii) *If* $\Lambda_1 = \infty$ *or if* $\Lambda_1 < \infty$ *and* $\sigma < \Phi(\Lambda_0, \Lambda_1)$, *equation* (6.2) *has no solution on* $[0, \delta)$.

(iii) *If* $\lambda_1 < \Lambda_1 < \infty$, $3\Lambda_0 \leq 2\Lambda_1$ *and* $\sigma \leq \Phi(\Lambda_0, \Lambda_1)$, *equation* (6.2) *has no solution on* $[0, \delta)$.

(iv) *If* $\sigma \geq 2\sqrt{\Lambda_1}$, *equation* (6.2) *has a solution* θ *on* $[0, \delta)$ *such that* $\theta(s) \geq \nu s$ *for all* $0 < s < \delta$, *where*

$$\nu := \frac{\sigma + \sqrt{\sigma^2 - 4\Lambda_1}}{2}. \tag{6.19}$$

Proof. (i) Suppose that (6.2) has a solution θ on $[0, \delta)$. Since c is positive, θ cannot be identically zero on $[0, \delta)$. On the other hand, $\theta(s) \leq \sigma s$ for all $0 < s < \delta$ by (6.2). This implies $\sigma > 0$. Furthermore, it implies that one can define $A := \sup\{\theta(s)/s : 0 < s < \delta\}$ in the knowledge that $0 < A \leq \sigma$. This means though that given any $0 < \varepsilon < A$ one can find an $s \in (0, \delta)$ such that

$$A - \varepsilon \leq \frac{\theta(s)}{s} = \sigma - \frac{1}{s}\int_0^s \frac{c(r)a'(r)}{\theta(r)}\,dr \leq \sigma - \frac{1}{s}\int_0^s \frac{c(r)a'(r)}{Ar}\,dr \leq \sigma - \frac{\Lambda_0}{A}.$$

Multiplying the above by $4A$ and letting $\varepsilon \downarrow 0$ yields $(2A - \sigma)^2 \leq \sigma^2 - 4\Lambda_0$. Whence, $\sigma > 0$, $\sigma \geq 2\sqrt{\Lambda_0}$ and $A \leq \mu$.

(ii) In the light of part (i) we only have to prove the assertion for $\Lambda_1 = \infty$ and $3\Lambda_0 < 2\Lambda_1 < \infty$. We note that (6.2) may be differentiated to yield

$$\theta'(s) = \sigma - \frac{c(s)a'(s)}{\theta(s)} \tag{6.20}$$

for almost all $0 < s < \delta$. Subsequently, multiplying by $2\theta(s)/s$,

$$\left(\frac{\theta^2}{s}\right)' + \frac{(2\sigma s - \mu s - \theta)(\mu s - \theta)}{s^2} + \frac{2ca'}{s} = \mu(2\sigma - \mu)$$

for almost all $0 < s < \delta$. Whence, noting that if (6.2) has a solution θ on $[0, \delta)$, there holds $\theta^2(s)/s \to 0$ as $s \downarrow 0$ by part (i), integrating the above identity from 0 to $s \in (0, \delta)$ and thereafter dividing by s we deduce

$$\begin{aligned} \mu(2\sigma - \mu) &= \frac{\theta^2(s)}{s^2} + \frac{1}{s}\int_0^s \frac{\{2\sigma r - \mu r - \theta(r)\}\{\mu r - \theta(r)\}}{r^2}\,dr \\ &\quad + \frac{2}{s}\int_0^s \frac{c(r)a'(r)}{r}\,dr \end{aligned} \tag{6.21}$$

for all $0 < s < \delta$. Recalling that $\sigma r \geq \mu r \geq \theta(r)$ for all $0 < r < \delta$ by part (i), this yields

$$\mu(2\sigma - \mu) \geq 2\Lambda_1. \tag{6.22}$$

This last inequality plainly cannot hold if $\Lambda_1 = \infty$, while elementary manipulation shows that, if $3\Lambda_0 \leq 2\Lambda_1 < \infty$, it is equivalent to $\sigma \geq \Phi(\Lambda_0, \Lambda_1)$.

(iii) We refine the proof of the previous part. If $\lambda_1 < \Lambda_1 < \infty$, there exists a $0 < s \le \delta$ such that the quantity on the last line of (6.21) actually equals Λ_1. Invoking Lemma 2.17 if necessary, for such an s the identity (6.21) subsequently yields (6.22) with strict inequality, from which the assertion follows.

(iv) It can be verified that the function $\theta_1(s) := \nu s$ is a solution of equation (1.9) on $[0,\delta)$ with $\sigma s + b(s)$ replaced by

$$\sigma_1 s + b_1(s) := \nu s + \int_0^s \frac{c(r)a'(r)}{\nu r}\, dr \le \left(\nu + \frac{\Lambda_1}{\nu}\right) s = \sigma s.$$

Subsequently the existence of a solution θ of (6.2) satisfying $\theta(s) \ge \theta_1(s)$ for all $s \in [0,\delta)$ follows from Lemma 2.18(ii) of the theory of the integral equation. □

As a consequence of the reaction-diffusion equation with a source term admitting a one-parameter family of semi-wavefront solutions with a given wave speed decreasing to 0 if it admits one at all, the investigation of the boundedness of the support of these waves is also more complicated than in the case of a sink term. It will be of help to this investigation to first consider the next lemma.

Lemma 6.7. *Consider equation* (1.9) *with two different wave speeds* σ_i*, sets of coefficients* a_i*,* b_i *and* c_i*, and corresponding solutions* θ_i *on* $[0,\delta)$ *for* $0 < \delta \le \ell$ *and* $i = 1,2$*. Suppose that the function* $s \mapsto \sigma_2 s + b_2(s) - \sigma_1 s - b_1(s)$ *is nondecreasing on* $[0,\delta)$*,* $c_2 a_2' \le c_1 a_1'$ *almost everywhere in* $(0,\delta)$*, and,* $c_1(s) \ne 0$ *or* $c_2(s) \ne 0$ *for all* $0 < s < \delta$ *Then, either* $\theta_2 \ge \theta_1$ *on* $[0,\delta)$*, or, there exists a* $\delta^* \in (0,\delta)$ *such that* $\theta_2 < \theta_1$ *on* $(0,\delta^*)$*.*

Proof. By Lemma 2.24, $\max\{\theta_1(s), \theta_2(s)\} > 0$ for any $0 < s < \delta$. Hence, if there exists an $s^* \in (0,\delta)$ such that $\theta_2(s^*) < \theta_1(s^*)$, by Lemma 4.7 there holds $\theta_2(s) < \theta_1(s)$ for all $0 < s < s^*$. This gives the result. □

The above lemma enables us to establish the following.

Lemma 6.8. *Let the assumptions of Lemma* 6.6 *hold.*

(i) *If* $\sigma > 2\sqrt{\Lambda_1}$*, then given any* $\gamma > \gamma^* := (\sigma - \sqrt{\sigma^2 - 4\Lambda_1})/2$ *equation* (6.2) *has at most one solution* θ *on* $[0,\delta)$ *such that*

$$\theta(s) \ge \gamma s \quad \text{for all } 0 < s < \delta. \tag{6.23}$$

(ii) *If for some* $A > 0$ *there holds* $c(s)a'(s) \ge As$ *for almost all* $0 < s < \delta$*, then given any solution* θ *of* (6.2) *on* $[0,\delta)$ *necessarily* $\sigma^2 \ge 4A$ *and*

$$\liminf_{s \downarrow 0} \frac{\theta(s)}{s} \ge \frac{\sigma - \sqrt{\sigma^2 - 4A}}{2}. \tag{6.24}$$

(iii) *If $\sigma > 2\sqrt{\Lambda_1}$, and, for some $\Lambda_1 < B \le \sigma^2/4$ there holds $c(s)a'(s) \le Bs$ for almost all $0 < s < \delta$, then given any solution θ of* (6.2) *on* $[0,\delta)$ *other than the maximal solution, necessarily*

$$\limsup_{s\downarrow 0} \frac{\theta(s)}{s} \le \frac{\sigma - \sqrt{\sigma^2 - 4B}}{2}. \tag{6.25}$$

(iv) *If ca' is absolutely continuous on $[0,\delta)$, $(ca')(0) = 0$, and, for some A there holds $(ca')'(s) \ge A$ for almost all $0 < s < \delta$, then given any solution θ of* (6.2) *on* $[0,\delta)$, *necessarily $\sigma^2 \ge 4A$ and*

$$\liminf_{s\downarrow 0} \frac{\theta(s)}{c(s)a'(s)} \ge \frac{2}{\sigma + \sqrt{\sigma^2 - 4A}}. \tag{6.26}$$

(v) *If $\sigma > 2\Lambda_1$, ca' is absolutely continuous on $[0,\delta)$, $(ca')(0) = 0$, and, for some $\Lambda_1 < B \le \sigma^2/4$ there holds $(ca')'(s) \le B$ for almost all $0 < s < \delta$, then given any solution θ of* (6.2) *on* $[0,\delta)$ *other than the maximal solution, necessarily*

$$\limsup_{s\downarrow 0} \frac{\theta(s)}{c(s)a'(s)} \le \frac{2}{\sigma + \sqrt{\sigma^2 - 4B}}. \tag{6.27}$$

Proof. (i) We let X denote the set of real functions ψ defined on $[0,\delta]$ such that $\nu \le \psi \le \sigma$, where ν is defined by (6.19), and define the mapping

$$F(\psi) := \sigma - \frac{1}{s}\int_0^s \frac{c(r)a'(r)}{r\psi(r)}\,dr$$

on X. This mapping can be shown to be a contraction on X. Therefore, by the Banach–Cacciopoli contraction mapping principle, F has a unique fixed point ψ in X. Setting $\theta = s\psi(s)$ subsequently gives the existence of a unique solution of (6.2) in the class of functions satisfying

$$\nu s \le \theta(s) \le \sigma s \quad \text{for } 0 \le s \le \delta. \tag{6.28}$$

However, by part (i) of Lemma 6.6 and Lemma 2.17 any solution of (6.2) must satisfy the right-hand inequality in (6.28). On the other hand, if a solution satisfies (6.23) for some $\nu > \gamma > \gamma^*$, then substituting (6.23) in the right-hand side of (6.2) we compute $\theta(s) \ge \gamma_1 s$ for all $0 < s < \delta$ where $\gamma_1 = \sigma - \Lambda_1/\gamma$. Next substituting this new inequality in the right-hand side of (6.2) we find $\theta(s) \ge \gamma_2 s$ for all $0 < s < \delta$ with $\gamma_2 = \sigma - \Lambda_1/\gamma_1$. Repeating this process delivers a sequence of values γ_k such that $\theta(s) \ge \gamma_k s$ for all $0 < s < \delta$ and $k \ge 1$. Moreover, this sequence is increasing, and such that $\gamma_k \to \nu$ as $k \to \infty$. Recalling Lemma 2.17, this implies that θ must also satisfy the left-hand inequality in (6.28). In summary then, any solution θ that satisfies (6.23) for some $\gamma > \gamma^*$ must satisfy (6.28), and in this class of functions (6.2) is uniquely solvable.

(ii) Lemma 6.6(i) implies that $\sigma^2 \geq 4A$. To prove (6.24) we may therefore suppose that $A < \sigma^2/4$ without loss of generality. Let $c_2(r)a_2'(r) := Ar$, and, observe that by Lemma 2.19(ii) for every γ equation (6.2) with $c_2(r)a_2'(r)$ in lieu of $c(r)a'(r)$ admits a unique solution θ_γ such that

$$\theta_\gamma(s) = \beta_1 s + \gamma s^{\beta_2/\beta_1} + \mathcal{O}(s^{(2\beta_2-\beta_1)/\beta_1}) \quad \text{as } s \downarrow 0, \tag{6.29}$$

where $\beta_1 := (\sigma - \sqrt{\sigma^2 - 4A})/2$ and $\beta_2 := (\sigma + \sqrt{\sigma^2 - 4A})/2$. Furthermore if $[0, \delta_\gamma)$ denotes its maximal interval of existence contained in $[0, \delta)$, θ_γ is positive on $(0, \delta_\gamma), \delta_\gamma$ depends continuously and monotonically on γ, and, $\delta_\gamma \to 0$ as $\gamma \to -\infty$. Simultaneously, $c_2(s)a_2'(s) \leq c(s)a'(s)$ for almost all $0 < s < \delta$, and, $c_2(s)a_2'(s) > 0$ for all $0 < s < \delta$. So, by Lemma 6.7, either

$$\theta(s) > \theta_\gamma(s) \quad \text{for all } 0 < s < \delta^* \quad \text{some } 0 < \delta^* < \delta_\gamma, \tag{6.30}$$

or, $\delta_\gamma = \delta$ and $\theta(s) \leq \theta_\gamma(s)$ for all $0 < s < \delta$. Consequently, since $\delta_\gamma \to 0$ as $\gamma \to -\infty$, we can choose a negative γ of sufficient magnitude that (6.30) holds. This gives (6.24).

(iii) Suppose to begin with that $\sigma^2 > 4B$. Set $\beta_1 := (\sigma - \sqrt{\sigma^2 - 4B})/2$ and $\beta_2 := (\sigma + \sqrt{\sigma^2 - 4B})/2$, and note that Lemma 2.19(ii) infers the existence of a solution θ_γ of equation (6.2) with $c(r)a'(r)$ replaced by $c_1(r)a_1'(r) := Br$ such that (6.29) holds for every γ. Furthermore, if $[0, \delta_\gamma)$ denotes its maximal interval of existence contained in $[0, \delta)$, θ_γ is positive on $(0, \delta_\gamma), \delta_\gamma$ is a continuous monotonic increasing function of γ, $\delta_\gamma = \delta$ for all $\gamma > 0$, and, $\theta_{\gamma'}(s) > \theta_\gamma(s)$ for all $0 < s < \delta_\gamma$ and $\gamma' > \gamma$. Finally, $c(s)a'(s) \leq c_1(s)a_1'(s)$ for almost all $0 < s < \delta$, and, $c(s) \neq 0$ for all $0 < s < \delta$. So, by Lemma 6.7, for any γ either

$$\theta(s) < \theta_\gamma(s) \quad \text{for all } 0 < s < \delta^* \quad \text{some } 0 < \delta^* < \delta_\gamma \tag{6.31}$$

or, $\theta(s) \geq \theta_\gamma(s)$ for all $0 < s < \delta_\gamma$. We deduce therefore that either (6.31) holds for some γ, or, $\theta(s) \geq \theta_\infty(s) := \sup\{\theta_\gamma(s) : 0 < \gamma < \infty\}$ for all $0 < s < \delta$. The function θ_∞ can be verified to be a solution of (6.2) with $c(r)a'(r)$ replaced by $c_1(r)a_1'(r)$ though. Whence, by Lemma 2.19(ii), $\theta_\infty(s) = \beta_2 s$ for all $0 < s < \delta$. However, by part (i) of the present lemma and Lemma 6.6(iv), the inequality $\theta(s) \geq \beta_2 s$ for $0 < s < \delta$ implies that θ must be the maximal solution of (6.2). Thus, since by hypothesis this is not the case, there has to be a γ such that (6.31) holds. This yields (6.25). If $\sigma^2 = 4B$ the result may be obtained similarly using Lemma 2.19(i) instead of Lemma 2.19(ii)

(iv) By part (ii) $\sigma^2 \geq 4A$. Choose $\gamma < 2/(\sigma + \sqrt{\sigma^2 - 4A})$. Observe that $\theta_2(s) := \gamma c(s)a'(s)$ is a solution of (1.9) on $[0, \delta)$ with $\sigma s + b(s)$ replaced by $\sigma_2 s + b_2(s) := \theta_2(s) + s/\gamma$. Furthermore, $s \mapsto \sigma_2 s + b_2(s) - \sigma s$ is nondecreasing on $[0, \delta)$. Hence, by Lemma 6.7, either

$$\theta(s) > \theta_2(s) \quad \text{for all } 0 < s < \delta^* \quad \text{some } 0 < \delta^* < \delta \tag{6.32}$$

or $\theta(s) \leq \theta_2(s)$ for all $0 < s < \delta$. However, substituting this last inequality in the right-hand side of (6.2) we compute $\theta(s) \leq (\sigma - 1/\gamma)s$ for all $0 < s < \delta$, which possibility is excluded by part (ii) of the present lemma. We are therefore forced to conclude that (6.32) must hold. The conclusion (6.26) follows.

(v) Note that the function $\theta_1(s) := \{2/(\sigma + \sqrt{\sigma^2 - 4B})\}c(s)a'(s)$ is a solution of (1.9) on $[0, \delta)$ with $\sigma s + b(s)$ replaced by $\sigma_1 s + b_1(s) := \theta_1(s) + \{(\sigma + \sqrt{\sigma^2 - 4B})/2\}s$. Furthermore, $s \mapsto \sigma s - \sigma_1 s - b_1(s)$ is nondecreasing on $[0, \delta)$. Hence, in this case by Lemma 6.7, either

$$\theta(s) < \theta_1(s) \quad \text{for all } 0 < s < \delta^* \quad \text{some } 0 < \delta^* < \delta \tag{6.33}$$

or $\theta(s) \geq \theta_1(s)$ for all $0 < s < \delta$. Substitution of the last inequality in the right-hand side of (6.2) yields $\theta(s) \geq \{(\sigma - \sqrt{\sigma^2 - 4B})/2\}s$ for all $0 < s < \delta$; whereafter, part (i) of the present lemma and Lemma 6.6(iv) imply that θ must be the maximal solution of (6.2). Thus, since θ is not the maximal solution, (6.33) must hold. This yields (6.27). □

Lemma 6.8 supplies the following.

Theorem 6.9 (Bounded support). *Suppose that the conditions of Theorem 6.5 hold. Fix $0 < \delta < \ell$.*

(a) *If*

$$\int_0^\delta \frac{a'(s)}{s}\, ds = \infty, \tag{6.34}$$

then every semi-wavefront solution decreasing to 0 is positive everywhere in its domain of definition.

(b) *If*

$$\int_0^\delta \frac{a'(s)}{s}\, ds < \infty, \tag{6.35}$$

then for every wave speed $\sigma > 2\sqrt{\lambda_1}$ there is a solution of this type whose support is bounded above. Moreover, if ca' is absolutely continuous on $[0, \delta)$, $(ca')(0) = 0$, $\lim\operatorname{ess\,sup}_{s\downarrow 0}(ca')'(s) \leq \lambda_1$, and,

$$\int_0^\delta \frac{1}{c(s)}\, ds = \infty,$$

then for every wave speed $\sigma > 2\sqrt{\lambda_1}$ there is exactly one distinct solution of this type whose support is bounded above and all other solutions of this type are positive everywhere in their domain of definition. On the other hand, if $\lim\operatorname{ess\,inf}_{s\downarrow 0} c(s)a'(s)/s > 0$, or, if ca' is absolutely continuous on $[0, \delta)$, $(ca')(0) = 0$, $\lim\operatorname{ess\,inf}_{s\downarrow 0}(ca')'(s) > -\infty$, and,

$$\int_0^\delta \frac{1}{c(s)}\, ds < \infty,$$

then for every wave speed $\sigma > 2\sqrt{\lambda_1}$ every solution of this type is such that its support is bounded above.

6.3 Smooth coefficients

Let us now turn to the case of equation (6.1) when ca' is continuously differentiable in $[0, \ell)$ and $(ca')(0) = 0$. The previous two sections already provide results should c be negative everywhere in $(0, \delta)$ or positive everywhere in $(0, \delta)$ for some $0 < \delta \leq \ell$. The outstanding situation is therefore that in which c vanishes at some point in $(0, \delta)$ for every $0 < \delta < \ell$. The following are our results for this case.

Theorem 6.10 (Existence). *Suppose that $ca' \in C^1([0, \ell))$, and, $c(u_i) = 0$ for a sequence of values $\{u_i\}_{i=1}^{\infty} \subset (0, \ell)$ such that $u_i \to 0$ as $i \to \infty$.*

(a) *If, for some $0 < \delta < \ell$ there holds*

$$\int_0^s c(u)a'(u)\,du < 0 \quad \text{for all } 0 < s \leq \delta, \tag{6.36}$$

then equation (6.1) has exactly one distinct semi-wavefront solution decreasing to 0 for every wave speed $\sigma \geq 0$ and no such solution for any wave speed $\sigma < 0$.

(b) *If, given any $0 < \delta < \ell$ there holds*

$$\int_0^s c(u)a'(u)\,du \geq 0 \quad \text{for some } 0 < s \leq \delta,$$

then equation (6.1) has exactly one distinct semi-wavefront solution decreasing to 0 for every wave speed $\sigma > 0$ and no such solution for any wave speed $\sigma \leq 0$.

Theorem 6.11 (Bounded support). *Suppose that the conditions of Theorem 6.10 hold. Fix $0 < \delta < \ell$. Let Q be defined by (6.3).*

(a) *If (6.34) holds, then every semi-wavefront decreasing to 0 is positive everywhere in its domain of definition.*

(b) *If (6.35) holds, then every solution of this type with wave speed $\sigma > 0$ is such that its support is bounded above. Moreover, if (6.36) and (6.4) hold, every solution of this type with wave speed $\sigma = 0$ is positive everywhere in its domain of definition. Whereas, if (6.36) and (6.5) hold, every solution of this type with wave speed $\sigma = 0$ is such that its support is bounded above.*

Both of these theorems follow from Lemma 2.25, Lemma 2.40 and the lemma below.

Lemma 6.12. *Suppose that the introductory conditions of Theorem* 6.10 *hold.*

(i) *If* $\sigma > 0$, *there exists a* $0 < \delta < \ell$ *such that equation* (6.2) *has a unique solution* θ *on* $[0, \delta)$. *Furthermore,*

$$\frac{\theta(s)}{s} \to \sigma \quad \text{as } s \downarrow 0. \tag{6.37}$$

(ii) *If* $\sigma = 0$, *then* (6.2) *has a solution if and only if* (6.36) *holds for some* $0 < \delta < \ell$, *in which event there is a unique solution* $\theta \equiv Q$ *on* $[0, \delta)$.

(iii) *If* $\sigma < 0$, *then* (6.2) *has no solution.*

Proof. We first show that (6.37) holds for any solution θ of equation (6.2) on an interval $[0, \delta)$ with $0 < \delta < \ell$. To do this, we observe that since $ca' \in C^1(0, \ell)$, θ must be positive on $(0, \delta)$ by Lemma 2.40. Moreover, in this light, (6.2) may be differentiated twice to deduce that $\theta \in C^2(0, \delta)$. To be specific, differentiating once gives (6.20) for all $0 < s < \delta$; while, multiplying (6.20) by $\theta(s)$ and differentiating again,

$$\theta(s)\theta''(s) = \{\sigma - \theta'(s)\}\theta'(s) - (ca')'(s) \quad \text{for } 0 < s < \delta. \tag{6.38}$$

Suppose now that

$$\liminf_{s \downarrow 0} \theta'(s) < \mu < \limsup_{s \downarrow 0} \theta'(s) \tag{6.39}$$

for some real number μ. Then by the regularity of θ there must be sequences $\{s_i^\pm\}_{i=1}^\infty \subset (0, \delta)$ such that $s_i^\pm \to 0$ as $i \to \infty$, $\theta'(s_i^\pm) = \mu$ and $\pm\theta''(s_i^\pm) \geq 0$ for all $i \geq 1$. Hence, noting that $(ca')'(s) \to 0$ as $s \downarrow 0$ by the hypotheses on ca', substituting $s = s_i^\pm$ in (6.38) and passing to the limit $i \to \infty$, we obtain $\pm(\sigma - \mu)\mu \geq 0$. This means that there are at most two values, namely $\mu = \sigma$ and $\mu = 0$, for which (6.39) might hold. On the other hand, since $\theta' \in C(0, \delta)$, the set of such values should comprise an open interval. This contradiction can only be resolved by the deduction that there are no numbers μ for which (6.39) holds. In other words, the two entities on the left- and right-hand side of (6.39) are equal. At the same time, from (6.20) we know that $\theta'(u_i) = \sigma$ for all $i \geq 1$ such that $u_i < \delta$. Taken together these conclusions imply that $\theta'(s) \to \sigma$ as $s \downarrow 0$. This gives (6.37) via l'Hôpital's rule. To proceed we distinguish the three cases in the statement of the lemma.

(i) Choose $0 < A < \sigma^2/4$ and thereafter $0 < \delta < \ell$ so small that $|(ca')'(s)| \leq A$ for all $0 \leq s \leq \delta$. Set $\iota := (\sigma - \sqrt{\sigma^2 - 4A})/2$. Then deploying the contraction-mapping argument used to prove part (i) of Lemma 6.8 it can be shown that (6.2) has a unique solution θ on $[0, \delta]$ such that $(\sigma - \iota)s \leq \theta(s) \leq (\sigma + \iota)s$ for all $0 \leq s \leq \delta$. Combining this deduction with (6.37) provides the desired result.

(ii) Multiplying (6.20) by $2\theta(s)$ and integrating, the integral equation is equivalent to

$$\theta^2(s) = -2 \int_0^s c(r)a'(r)\, dr.$$

This readily justifies the conclusion of the lemma.

(iii) This conclusion follows immediately from our deduction that (6.37) must hold for every solution of (6.2). □

To close this chapter, let us summarize the results we have obtained for the semilinear equation

$$u_t = u_{xx} + c(u) \tag{6.40}$$

with a smooth reaction term c, which has been the subject of great interest in the past.

Corollary 6.13 (of Theorems 6.1, 6.5 and 6.10). *Suppose that c is differentiable in I. Then every semi-wavefront solution of equation (6.40) decreasing to 0 is necessarily positive everywhere in its domain of definition. Moreover, the following is the case.*

(a) *If $c(u) < 0$ for all $0 < u < \delta$ for some $0 < \delta < \ell$, then the equation has exactly one distinct semi-wavefront solution decreasing to 0 for every wave speed σ.*

(b) *If*

$$\int_0^s c(r)\,dr < 0 \quad \text{for all } 0 < s < \delta$$

for some $0 < \delta < \ell$, and, given any $0 < \delta < \ell$ there exists a $0 < u < \delta$ such that $c(u) \geq 0$, then the equation has exactly one distinct semi-wavefront solution decreasing to 0 for every wave speed $\sigma \geq 0$, and, no such solution for any $\sigma < 0$.

(c) *If given any $0 < \delta < \ell$ there exists a $0 < u < \delta$ such that $c(u) \leq 0$ and a $0 < s < \delta$ such that*

$$\int_0^s c(r)\,dr \geq 0,$$

then the equation has exactly one distinct semi-wavefront solution decreasing to 0 for every wave speed $\sigma > 0$, and, no such solution for any $\sigma \leq 0$.

(d) *If $c(u) > 0$ for all $0 < u < \delta$ for some $0 < \delta < \ell$, and, $c'(0) = 0$, then the equation has a one-parameter family of distinct semi-wavefront solutions decreasing to 0 for every wave speed $\sigma > 0$ and no such solution for any $\sigma \leq 0$.*

(e) *If $c'(0) > 0$, then the equation has a one-parameter family of distinct semi-wavefront solutions decreasing to 0 for every wave speed $\sigma > 2\sqrt{c'(0)}$ and no such solution for any $\sigma < 2\sqrt{c'(0)}$. For $\sigma = 2\sqrt{c'(0)}$ the equation either has a one-parameter family of distinct solutions or no solution, whereby a sufficient condition for existence is $c(u) \leq c'(0)u$ for all $0 < u < \delta$ for some $0 < \delta < \ell$.*

Proof. In case (a) the existence of a distinct semi-wavefront for every σ is given by Theorem 6.1; while, noting that the variable Q defined by (6.3) is such that $Q(s)/s \to \sqrt{|c'(0)|}$ as $s \downarrow 0$, the positivity of the travelling wave is given by part (i) of Theorem 6.2(a). Cases (b) and (c) follow immediately from Theorems 6.10 and 6.11(a). With regard to the remaining cases, one can compute that the parameters defined by (6.14) and (6.15) take the values $\lambda_1 = \lambda_0 = c'(0)$. Furthermore, if $c(u) \leq c'(0)u$ for all $0 < u < \delta$ for some $0 < \delta < \ell$, the parameter Λ_1 defined by (6.16) is also equal to $c'(0)$. Cases (d) and (e) are subsequently a straightforward consequence of Theorems 6.5 and 6.9(a). □

Equation (6.17) provides an explicit example to show that in the case (e) above when the condition $c(u) \leq c'(0)u$ for all $0 < u < \delta$ for some $0 < \delta < \ell$ does not hold, there may or may not be solutions with the wave speed $\sigma = 2\sqrt{c'(0)}$. As discussed in Section 6.2, if $k \leq 1/4$ equation (6.17) has a one-parameter family of semi-wavefront solutions with the critical speed $2\sqrt{c'(0)}$ decreasing to 0, whereas if $k > 1/4$ it has no solutions with this wave speed.

Bibliographic notes

The most well-known equations of the class (6.1) are the *Fisher equation* or *logistic equation* [105, 294], the *Newell–Whitehead equation* or *amplitude equation* [216, 217], the *Zeldovich equation* [78], the *KPP equation* [186], and, the *Nagumo equation* or *bistable equation* [56, 201, 211, 213, 214]. The importance of this class has been highlighted in several general surveys [56, 78, 82, 101, 210, 211, 222, 279].

For the semilinear version of equation (6.1) with a negative reaction term, the significance of the quantity Q defined by (6.3) was recognized as early as 1938 by Zel'dovich and Frank-Kamenetskii [314, 318].

For further details of the specific occurrence of equation (6.13) as a model of the diffusion of oxygen in absorbing tissue we refer to [73, 75, 76, 96], while its origin as a reformulation of the Stefan problem is described in [75, 92, 96, 264].

Studies of travelling-wave solutions of the semilinear version (6.23) of equation (6.1) with a smooth reaction term can be found in [20, 21, 25, 102, 103, 134, 142, 143, 149, 150, 211, 296, 297].

The reference [199] and the reference [110] may be viewed as containing early forerunners, obtained with varying degrees of rigour, to Theorems 6.1 and 6.2, and to Theorems 6.5 and 6.9 respectively.

The proof of Lemma 6.6 part (ii) and of Lemma 6.8 part (i) extends phase-plane arguments for particular examples of equation (6.1) with a source term, previously employed by Atkinson et al. [22], to integral equation arguments for general equations of this class. The proof of Lemma 6.12 part (i) relies on an adaption of an argument for the semilinear equation (6.40) in [297] to fit the more general equation (6.1).

Chapter 7

Power-law equations

We turn now to the application of the integral equation (1.9) for the definitive analysis of semi-wavefront solutions for two specific classes of equation (1.1). The first of these is

$$u_t = (u^m)_{xx} + b_0 (u^n)_x + \begin{cases} c_0 u^p & \text{for } u > 0 \\ 0 & \text{for } u = 0 \end{cases} \tag{7.1}$$

where m, n, p, b_0 and c_0 are real parameters. Equations of this type have long been of interest as a tractable prototype for more general equations of the class (1.1). Without the convection and reaction terms, equation (7.1) is simply the porous media equation. The integral equation approach leads to the following characterization of travelling-wave solutions for this category of equations.

Theorem 7.1. *Let $m > 0$, $n > 0$, $p + m > 0$, b_0 and c_0 be real constants.*

(a) *Suppose that $c_0 < 0$. Then for every wave speed σ equation (7.1) has exactly one distinct semi-wavefront solution decreasing to 0. Moreover, the support of this solution is bounded above if and only if the value q in Table 7.1 satisfies the condition $q < m$.*

(b) *Suppose that $c_0 = 0$. Then for every wave speed σ equation (7.1) has at most one distinct semi-wavefront solution decreasing to 0. There is such a solution if and only if there is a value q in Table 7.2. Moreover, the support of this solution is bounded above if and only if $q < m$.*

(c) *Suppose that $c_0 > 0$. Then for every wave speed σ equation (7.1) has either a one-parameter family of distinct semi-wavefront solution decreasing to 0 in the sense of Definition 4.5 or no such solution. There are such solutions if and only if there is a value q in Table 7.3, where*

$$b^* := 2\sqrt{\frac{mc_0}{n}} \quad \textit{and} \quad \sigma^* = 2\sqrt{mc_0}.$$

Moreover, there is such a solution whose support is bounded above if and only if $q < m$.

Proof. According to the theory we have already developed in this monograph, an equation of the class (1.1) admits a semi-wavefront solution with wave speed σ decreasing to 0 if and only if the integral equation (1.9) admits a solution satisfying the integrability condition. Moreover, there is a decreasing semi-wavefront solution

n	$m+p$	b_0	wave speed	q
$n<1$	$m+p>2n$	$b_0>0$	all σ	n
		$b_0<0$	all σ	$m+p-n$
	$m+p\leq 2n$	any b_0	all σ	$(m+p)/2$
$n=1$	$m+p>2$	any b_0	$\sigma>-b_0$	1
			$\sigma=-b_0$	$(m+p)/2$
			$\sigma<-b_0$	$m+p-1$
	$m+p\leq 2$	any b_0	all σ	$(m+p)/2$
$n>1$	$m+p>2n$	$b_0>0$	$\sigma>0$	1
			$\sigma=0$	n
			$\sigma<0$	$m+p-1$
		$b_0<0$	$\sigma>0$	1
			$\sigma=0$	$m+p-n$
			$\sigma<0$	$m+p-1$
	$2n\geq m+p>2$	any b_0	$\sigma>0$	1
			$\sigma=0$	$(m+p)/2$
			$\sigma<0$	$m+p-1$
	$m+p\leq 2$	any b_0	all σ	$(m+p)/2$

Table 7.1: Value of q for which the solution of (7.2) or (7.5) with $c_0<0$ satisfies (7.3) or (7.6) respectively.

with bounded support if and only if (1.9) has a solution θ on some interval $[0,\delta)$ for which (2.35) holds. Consequently, recalling that even if equation (1.9) admits more than one solution, there is one which is maximal; to determine whether or not (1.9) has a solution satisfying the integrability condition and possibly also (2.35), it suffices to investigate the existence and behaviour of a maximal solution of this equation. For the partial differential equation (7.1) the integral equation (1.9) reads

$$\theta(s)=\sigma s+b_0 s^n-mc_0\int_0^s\frac{r^{m+p-1}}{\theta(r)}\,dr. \tag{7.2}$$

We assert that the maximal solution θ of (7.2) is necessarily such that

$$\theta(s)\sim\theta_0 s^q \quad \text{as } s\downarrow 0 \quad \text{for some } \theta_0>0 \tag{7.3}$$

and $q>0$ as stated in Tables 7.1 – 7.3. In this event, the number of semi-wavefront solutions of equation (7.1) decreasing to 0 is given directly by Theorems 2.9, 4.4 and 4.6. Furthermore, by substitution in (2.35), the existence of a semi-wavefront solution with bounded support is determined by the simple criterion $q<m$. To proceed, we define

$$Q(s):=\sqrt{\tfrac{2m}{m+p}\,|c_0|}\;s^{(m+p)/2}$$

n	b_0	wave speed	q
$n<1$	$b_0>0$	all σ	n
	$b_0<0$	all σ	none
$n=1$	any b_0	$\sigma>-b_0$	1
		$\sigma\le -b_0$	none
$n>1$	$b_0>0$	$\sigma>0$	1
		$\sigma=0$	n
		$\sigma<0$	none
	$b_0\le 0$	$\sigma>0$	1
		$\sigma\le 0$	none

Table 7.2: Value of q for which the solution of (7.2) or (7.5) with $c_0=0$ satisfies (7.3) or (7.6) respectively.

and

$$\varkappa := \lim_{s\downarrow 0} \frac{\sigma s + b_0 s^n}{Q(s)},$$

and distinguish the three cases in the statement of the theorem.

(a) When $c_0<0$ the existence of a unique solution θ of (7.2) on $[0,\infty)$ is given by Lemma 2.16(i). Moreover, when $\varkappa>-\infty$ its behaviour as $s\downarrow 0$ is given by Lemma 2.16(ii). Regarding its behaviour when $\varkappa=-\infty$, consider equation (7.2) with $\sigma_i s + b^{(i)}(s) := \{1-(-1)^i\varepsilon\}(\sigma s + b_0 s^n) + \theta_i(s)$, where

$$\theta_i(s) := \frac{mc_0 s^{m+p}}{\{1-(-1)^i\varepsilon\}(\sigma s + b_0 s^n)},$$

$0<\varepsilon<1$ and $i=1,2$, in lieu of $\sigma s + b_0 s^n$. It can be verified that θ_i is a solution of this auxiliary equation for $i=1,2$. Furthermore, in a small enough interval $[0,\delta)$ the functions $s\mapsto \sigma_2 s + b^{(2)}(s) - \sigma s + b_0 s^n$ and $s\mapsto \sigma s + b_0 s^n - \sigma_1 s - b^{(1)}(s)$ are both nondecreasing. Subsequently Lemma 2.18(i) implies that $\theta_1\le\theta\le\theta_2$ on $[0,\delta)$. The behaviour of θ as $s\downarrow 0$ follows by passing to the limit $\varepsilon\downarrow 0$.

(b) When $c_0=0$ the integral equation (7.2) reduces to the simple identity $\theta(s)=\sigma s + b_0 s^n$. Subsequently, it is easy to check whether or not there is a 'solution' satisfying the integrability condition and what its behaviour as $s\downarrow 0$ is. See Chapter 5 for further particulars of the principles involved.

(c) When $c_0>0$ Lemma 2.15 parts (i) and (ii) imply that equation (7.2) has a solution if and only if $\varkappa\ge 2$. Moreover, when $\varkappa=\infty$ the behaviour of the maximal solution θ of equation (7.2) as $s\downarrow 0$ follows from Lemma 2.15(i). To obtain the corresponding behaviour of the maximal solution θ when $2<\varkappa<\infty$, consider (7.2) with $\sigma s + b_0 s^n$ replaced by $\{\varkappa+(-1)^i\varepsilon\}Q(s)$ for some $0<\varepsilon<\varkappa-2$ and $i=1,2$. By Lemma 2.19 part (ii) this auxiliary equation

n	$m+p$	b_0	wave speed	q
$n<1$	$m+p>2n$	$b_0>0$	all σ	n
		$b_0<0$	all σ	none
	$m+p=2n$	$b_0\geq b^*$	all σ	n
		$b_0<b^*$	all σ	none
	$m+p<2n$	any b_0	all σ	none
$n=1$	$m+p>2$	any b_0	$\sigma>-b_0$	1
			$\sigma\leq -b_0$	none
	$m+p=2$	any b_0	$\sigma\geq\sigma^*-b_0$	1
			$\sigma<\sigma^*-b_0$	none
	$m+p<2$	any b_0	all σ	none
$n>1$	$m+p>2n$	$b_0>0$	$\sigma>0$	1
			$\sigma=0$	n
			$\sigma<0$	none
		$b_0\leq 0$	$\sigma>0$	1
			$\sigma\leq 0$	none
	$m+p=2n$	$b_0\geq b^*$	$\sigma>0$	1
			$\sigma=0$	n
			$\sigma<0$	none
		$b_0<b^*$	$\sigma>0$	1
			$\sigma\leq 0$	none
	$2n>m+p>2$	any b_0	$\sigma>0$	1
			$\sigma\leq 0$	none
	$m+p=2$	any b_0	$\sigma\geq\sigma^*$	1
			$\sigma<\sigma^*$	none
	$m+p<2$	any b_0	all σ	none

Table 7.3: Value of q for which the maximal solution of (7.2) with $c_0>0$ satisfies (7.3).

has a maximal solution $\theta_i := ([\varkappa+(-1)^i\varepsilon+\sqrt{\{\varkappa+(-1)^i\varepsilon\}^2-4}]/2)Q(s)$; while observing that $(\varkappa+\varepsilon)Q(s)\geq\sigma s+b_0s^n\geq(\varkappa-\varepsilon)Q(s)$ in a small enough interval $[0,\delta)$, by Lemma 2.18(ii) there holds $\theta_1\leq\theta\leq\theta_2$ on $[0,\delta)$ for some $\delta>0$. The required relation (7.3) follows by letting $\varepsilon\downarrow 0$. For the remaining case, $\varkappa=2$, an upper estimate of the behaviour of the maximal solution θ of equation (7.2) as $s\downarrow 0$ may be obtained by following the above argument. The corresponding lower estimate is given by Lemma 2.15(i). □

The second class of equations that we consider in this chapter are those of the form

$$u_t=(u\,|\ln u|^{-m})_{xx}+b_0(u\,|\ln u|^{1-n})_x+\begin{cases} c_0u\,|\ln u|^{2-p} & \text{for } u>0\\ 0 & \text{for } u=0\end{cases} \tag{7.4}$$

where m, n, p, b_0 and c_0 are real parameters. Equations of this type fall into the class (1.1) if we take the interval $I = [0, \ell]$ with $0 < \ell < 1$ and implicitly define the coefficients for $u = 0$ by continuity. Equations of this type constitute a weak perturbation of the linear version of (1.1).

Theorem 7.2. *Let m, n, p, b_0 and c_0 be real constants.*

(a) *Suppose that $c_0 < 0$. Then for every wave speed σ equation (7.4) has exactly one distinct semi-wavefront solution decreasing to 0. Moreover, the support of this solution is bounded above if and only if the value q in Table 7.1 satisfies the condition $q < m$.*

(b) *Suppose that $c_0 = 0$. Then for every wave speed σ equation (7.4) has at most one distinct semi-wavefront solution decreasing to 0. There is such a solution if and only if there is a value q in Table 7.2. Moreover, the support of this solution is bounded above if and only if $q < m$.*

(c) *Suppose that $c_0 > 0$. Then for every wave speed σ equation (7.4) has either a one-parameter family of distinct semi-wavefront solution decreasing to 0 or no such solution. There are such solutions if and only if there is a value q in Table 7.3 with the corrections in Table 7.4, where*

$$b^* := \sigma^* = 2\sqrt{c_0}.$$

Moreover, there is such a solution whose support is bounded above if and only if $q < m$.

Proof. The proof of this theorem is similar to that of the previous one. For the partial differential equation (7.4) the integral equation (1.9) becomes

$$\theta(s) = \sigma s + b_0 s\, |\ln s|^{1-n} - c_0 \int_0^s \frac{r\, |\ln r|^{2-m-p} \left(1 + m\, |\ln r|^{-1}\right)}{\theta(r)}\, dr. \tag{7.5}$$

The corresponding behaviour asserted for the maximal solution θ of this equation is

$$\theta(s) \sim \theta_0 s\, |\ln s|^{1\ \ q} \quad \text{as } s \downarrow 0 \quad \text{for some } \theta_0 > 0 \tag{7.6}$$

and $q > 0$. The appropriate variable which has to be examined to determine this is

$$\begin{aligned} Q(s) \;&:=\; \sqrt{|c_0|}\, s\, |\ln s|^{1-(m+p)/2} \\ &\quad \times \left\{ 1 + (m - p + 2) s^{-2}\, |\ln s|^{m+p-2} \int_0^s r\, |\ln r|^{1-m-p}\, dr \right\}^{1/2} \end{aligned}$$

with corresponding limit $\varkappa := \lim_{s \downarrow 0} (\sigma s + b_0 s\, |\ln s|^{1-n}) / Q(s)$. The proof of the theorem may subsequently be completed following the arguments used above. To be specific, everything is analogous to the proof of Theorem 7.1 except for the

n	$m+p$	b_0	adaptation	wave speed	q
$n<-1$	$m+p=2n$	$b_0=b^*$	$m>n-1$	all σ	none
$n=-1$	$m+p=2n$	$b_0=b^*$	$m>-2$	all σ	none
			$m=-2$	$\sigma\geq-\sqrt{c_0}/4$	n
				$\sigma<-\sqrt{c_0}/4$	none
$-1<n<0$	$m+p=2n$	$b_0=b^*$	$m>n-1$	all σ	none
			$m=n-1$	$\sigma\geq 0$	n
				$\sigma<0$	none
$n=0$	$m+p=2n$	$b_0=b^*$	any m	$\sigma\geq(m+1)\sqrt{c_0}$	n
				$\sigma<(m+1)\sqrt{c_0}$	none
$0<n<1$	$m+p=2n$	$b_0=b^*$	$m>n-1$	$\sigma>0$	n
				$\sigma\leq 0$	none
			$m\leq n-1$	$\sigma\geq 0$	n
				$\sigma<0$	none
$n=1$	$m+p=2$	any b_0	$m>0$	$\sigma>\sigma^*-b_0$	1
				$\sigma\leq\sigma^*-b_0$	none
$1<n<2$	$m+p=2n$	$b_0=b^*$	$m>n-1$	$\sigma>0$	1
				$\sigma\leq 0$	none
	$m+p=2$	$b_0<0$	any m	$\sigma>\sigma^*$	1
				$\sigma\leq\sigma^*$	none
$n=2$	$m+p=2n$	$b_0=b^*$	$m>1$	$\sigma>0$	1
				$\sigma\leq 0$	none
	$m+p=2$	any b_0	$m>b_0/\sqrt{c_0}$	$\sigma>\sigma^*$	1
				$\sigma\leq\sigma^*$	none
$2<n<3$	$m+p=2n$	$b_0=b^*$	$m>n-1$	$\sigma>0$	1
				$\sigma\leq 0$	none
	$m+p=2$	$b_0\geq 0$	$m>0$	$\sigma>\sigma^*$	1
				$\sigma\leq\sigma^*$	none
		$b_0<0$	$m\geq 0$	$\sigma>\sigma^*$	1
				$\sigma\leq\sigma^*$	none
$n=3$	$m+p=2n$	$b_0=b^*$	$m>2$	$\sigma>0$	1
				$\sigma\leq 0$	none
	$m+p=2$	$b_0\geq-\sqrt{c_0}/4$	$m>0$	$\sigma>\sigma^*$	1
				$\sigma\leq\sigma^*$	none
		$b_0<-\sqrt{c_0}/4$	$m\geq 0$	$\sigma>\sigma^*$	1
				$\sigma\leq\sigma^*$	none
$n>3$	$m+p=2n$	$b_0=b^*$	$m>n-1$	$\sigma>0$	1
				$\sigma\leq 0$	none
	$m+p=2$	any b_0	$m>0$	$\sigma>\sigma^*$	1
				$\sigma\leq\sigma^*$	none

Table 7.4: Exceptions to Table 7.3 displaying the value of q for which the maximal solution of (7.5) with $c_0>0$ satisfies (7.6).

existence result in the case $c_0 > 0$ and $\varkappa = 2$. In this case, the calculations are more subtle. When $c_0 > 0$ and $\varkappa = 2$, Lemma 2.15 infers that the integral equation (7.5) admits a solution if and only if $\mu \geq -1/4$, where $\mu := \lim_{s\downarrow 0}\{(\sigma s + b_0 s\,|\ln s|^{1-n})/Q(s) - 2\}\,|\ln Q(s)|^2$. $\square$

We note that when $b_0 = 0$ the value of n in equations (7.1) and (7.4) is immaterial. To prevent any confusion, this possibility has therefore only been included in Tables 7.1 – 7.4 under those values of n and $m + p$ where this leads to no ambiguity. The reader actively wishing to use the tables, should search for $b_0 = 0$ under $n = 1$.

Theorem 7.2 brings a mistake in [127] to light. Consider an arbitrary solution u of an equation of the class (1.1) in the strip $S := (-\infty, \infty) \times (0, T]$ or the half-strip $H := (0, \infty) \times (0, T]$ with $0 < T < \infty$. Assume that u is continuous in $\overline{H}$ and define $\zeta(t) := \sup\{x \in (0, \infty) : u(x, t) > 0\}$ for all $0 \leq t \leq T$ with the convention that $\zeta(t) = 0$ if the supremum is taken over an empty set. Then the equation is said to display *finite speed of propagation* if $0 < \zeta(0) < \infty$ infers that $\zeta(t) < \infty$ for all $0 < t \leq T$. A similar definition can be made for a solution u defined in the strip S or the half-strip $(-\infty, 0) \times (0, T]$ with $\zeta(t) := \inf\{x \in (-\infty, 0) : u(x, t) > 0\}$. However, since we are considering a general equation of the class (1.1), by the simple expedient of changing the variable $x \mapsto -x$ there is no loss of generality in only considering the first alternative. The porous media equation, $u_t = (u^m)_{xx}$ with $m > 1$, and the linear heat equation, $u_t = u_{xx}$, are the archetypal examples of equations which do and do not display finite speed of propagation respectively. Then in [124, 127] it was effectively shown that an equation of the class (1.1) displays finite speed of propagation if and only if it admits a semi-wavefront solution whose support is bounded above. Among other applications of this result, the parameter values for which equation (7.1) and for which equation (7.4) with a slightly different notation display finite speed of propagation were presented. From Theorem 7.2 it is clear that the conclusions presented in [127] for (7.4) are not entirely correct. We give the true result below as a corollary of Theorem 7.2.

Corollary 7.3 (of Theorem 7.2). *Equation* (7.4) *with real parameters m, n, p, b_0 and c_0 admits finite speed of propagation if and only if one of the following hold.*

(i) $c_0 < 0$, $n \geq 1$ *or* $b_0 = 0$, *and* $m > \min\{p, 1\}$.

(ii) $c_0 < 0$, $n < 1$, $b_0 < 0$ *and* $p < \min\{m, n\}$.

(iii) $c_0 < 0$, $n < 1$, $b_0 > 0$, *and* $m > \min\{n, p\}$.

(iv) $c_0 = 0$, $n \geq 1$ *or* $b_0 = 0$, *and* $m > 1$.

(v) $c_0 = 0$, $n < 1$, $b_0 > 0$ *and* $m > n$.

(vi) $c_0 > 0$, $n \geq 1$ *or* $b_0 = 0$, $m > 1$ *and* $m + p \geq 2$.

(vii) $c_0 > 0$, $0 \leq n < 1$, $0 < b_0 < 2\sqrt{c_0}$, $m > n$ *and* $m + p > 2n$.

(viii) $c_0 > 0$, $0 \le n < 1$, $b_0 \ge 2\sqrt{c_0}$, $m > n$ *and* $m + p \ge 2n$.

(ix) $c_0 > 0$, $n < 0$, $0 < b_0 \le 2\sqrt{c_0}$, $m > n$ *and* $m + p > 2n$.

(x) $c_0 > 0$, $n < 0$, $b_0 > 2\sqrt{c_0}$, $m > n$ *and* $m + p \ge 2n$.

Proof. From Tables 7.1 – 7.4 it follows that the integral equation (7.5) admits a solution θ for large values of σ if and only if one of the following ten combinations hold. Moreover, if σ is sufficiently large, then (7.6) holds with the value of q stated below. The result is subsequently obtained by recalling that the partial differential equation (7.4) admits finite speed of propagation if and only if for a large enough wave speed σ the equation has a semi-wavefront solution whose support is bounded above, while, by Theorem 7.2 such a travelling wave exists if and only if the appropriate value of q is less than m.

(i) $c_0 < 0$, $n \ge 1$ or $b_0 = 0$; with $q = \min\{(m+p)/2, 1\}$.

(ii) $c_0 < 0$, $n < 1$, $b_0 < 0$; with $q = \max\{m+p-n, (m+p)/2\}$.

(iii) $c_0 < 0$, $n < 1$, $b_0 > 0$; with $q = \min\{n, (m+p)/2\}$.

(iv) $c_0 = 0$, $n \ge 1$ or $b_0 = 0$; with $q = 1$.

(v) $c_0 = 0$, $n < 1$, $b_0 > 0$; with $q = n$.

(vi) $c_0 > 0$, $n \ge 1$ or $b_0 = 0$, $m + p \ge 2$; with $q = 1$.

(vii) $c_0 > 0$, $\min\{m+1, 0\} \le n < 1$, $0 < b_0 < 2\sqrt{c_0}$, $m + p > 2n$; with $q = n$.

(viii) $c_0 > 0$, $\min\{m+1, 0\} \le n < 1$, $b_0 \ge 2\sqrt{c_0}$, $m + p \ge 2n$; with $q = n$.

(ix) $c_0 > 0$, $n < \min\{m+1, 0\}$, $0 < b_0 \le 2\sqrt{c_0}$, $m + p > 2n$; with $q = n$.

(x) $c_0 > 0$, $n < \min\{m+1, 0\}$, $b_0 > 2\sqrt{c_0}$, $m + p \ge 2n$; with $q = n$. □

This rectifies Lemma 14 in [127] where the mistake occurs in not recognizing the distinction between the cases $\min\{m+1, 0\} \le n < 1$ and $n < \min\{m+1, 0\}$ for $c_0 > 0$, and in subscribing the conclusions of (ix) and (x) to all $c_0 > 0$, $n < 1$.

Bibliographic notes

The *porous media equation*, $u_t = (u^m)_{xx}$, has acquired its name because of its description of the flow of an adiabatic gas in a porous medium [10, 193, 212]. The equation also arises in other contexts [10, 118]. It can be used to describe nonlinear heat transfer [316, 317], concentration-dependent diffusion [74], the motion of plasma particles in a magnetic field [46], and the evolution of biological populations [140, 210, 211]. In the case $m = 2$ the equation can be found in boundary-layer theory [265], and as a dimensionless reformulation of the Boussinesq equation in hydrology [30]. In the case $m = 7/2$ it arises in the study of solar prominences [248], while the case $m = 3/2$ occurs as a model of the imbibition of a wetting liquid in a capillary tube [84]. The equation has even been proposed as a suitable model for the spread of intergalactic civilizations [134, 220]. For overviews of the mathematical theory of the equation, see [16, 17, 171, 235, 289].

The development of the mathematical theory for the porous media equation with convection $u_t = (u^m)_{xx} + (u^n)_x$, where $m > 0$ and $n > 0$ can be traced through [8, 9, 72, 80, 81, 116, 117, 119–121, 125, 129, 130]. Interestingly, semi-wavefront solutions have played an important part throughout this development. For information on the theory for the porous media equation with absorption $u_t = (u^m)_{xx} - u^p$ or with a source term $u_t = (u^m)_{xx} + u^p$, where $m > 0$ and $p > 0$ one may refer to [6, 160, 182, 229, 287]. The so-called quenching problem, $u_t = (u^m)_{xx} - u^{-p}$ as it were, with $m > 0$ and $p > 0$ is dealt with in [178, 239]. For the full equation (7.1), existence, uniqueness and related results have been proven in [274].

For further details concerning the interest in equation (7.1) and the role it has played in the development of a theory for equations of the class (1.1), see [127] and the many references cited therein.

Chapter 8

Wavefronts

The subject of this and the ensuing two chapters is the existence of wavefront solutions of equations of the class (1.1). Thus we shall be concerned with solutions of equation (1.1) of the form $u = f(x - \sigma t)$ where $f(\xi)$ is defined and monotonic for $-\infty < \xi < \infty$, and where $f(\xi) \to \ell^{\pm}$ as $\xi \to \pm\infty$ for some $\ell^{\pm} \in I$ for which $c(\ell^{\pm}) = 0$ and $\ell^{+} \neq \ell^{-}$. Such solutions connecting two equilibrium states of the equation have long been of interest. Without any loss of generality we shall take $\ell^{-} = \ell < \infty$ and $\ell^{+} = 0$, and, in line with Definition 2.4, term the solution a wavefront from ℓ to 0.

8.1 Admissible wave speeds

Our first result states that the set of speeds for which wavefront solutions exist is connected.

Theorem 8.1. *Suppose that $\ell < \infty$. Then if equation* (1.1) *has a wavefront solution from ℓ to 0 with speed σ_1 and with speed $\sigma_2 > \sigma_1$ the same can be said for all $\sigma_1 \leq \sigma \leq \sigma_2$.*

Proof. If equation (1.1) has a wavefront solution with wave speed σ_1 from ℓ to 0, then by Theorem 2.10 the integral equation (1.9) with $\sigma = \sigma_1$ has a solution satisfying the integrability condition on $[0, \ell]$; while if equation (1.1) has a wavefront solution with wave speed σ_2 from ℓ to 0, then by Theorem 2.10 and Lemma 3.5 the integral equation

$$\Theta(s) = -\sigma s + b(\ell - s) - b(\ell) + \int_0^s \frac{c(\ell - r)a'(\ell - r)}{\Theta(r)}\, dr \tag{8.1}$$

with $\sigma = \sigma_2$ has a solution satisfying the integrability condition on $[0, \ell]$. Subsequently, by Lemma 2.18(i) from the theory of the integral equation, (1.9) has a solution satisfying the integrability condition on $[0, \ell]$ for all $\sigma \geq \sigma_1$, and equation (8.1) has a solution satisfying the integrability condition on $[0, \ell]$ for all $\sigma \leq \sigma_2$. Lemma 3.5 and Theorem 2.10 then provide the required result. $\square$

It follows from Theorem 8.1 that the set of wave speeds for which equation (1.1) has a wavefront solution from ℓ to 0 is either empty, a single value, or an interval. The next theorem provides a characterization in the form of a 'variational principle'.

Theorem 8.2. *Suppose that $\ell < \infty$. Let $\mathcal{R}$ denote the set of nonnegative continuous functions ψ defined on I such that*

$$\int_{s_0}^{s_1} \frac{\{1+|c(r)|\}a'(r)}{\psi(r)}\,dr < \infty \quad \text{for all } 0 < s_0 < s_1 < \ell, \tag{8.2}$$

let $\mathcal{S}_0$ denote the subset of functions in $\mathcal{R}$ such that $\psi(0) = 0$ and

$$\lim_{\varepsilon \downarrow 0} \int_{\varepsilon}^{s} \frac{c(r)a'(r)}{\psi(r)}\,dr \quad \text{exists and is finite for every } 0 < s < \ell,$$

and let $\mathcal{S}_1$ denote the subset of functions in $\mathcal{R}$ such that $\psi(\ell) = 0$ and

$$\lim_{\varepsilon \downarrow 0} \int_{s}^{\ell-\varepsilon} \frac{c(r)a'(r)}{\psi(r)}\,dr \quad \text{exists and is finite for all } 0 < s < \ell.$$

Define the functionals

$$\mathcal{F}_s(\psi) := \sup_{0<r<s<\ell} \left\{ \frac{\psi(s) - \psi(r) - b(s) + b(r) + \int_r^s \frac{c(w)a'(w)}{\psi(w)}\,dw}{s-r} \right\}$$

and $\mathcal{F}_i$ similarly with "sup" replaced by "inf",

$$\mathcal{G}_s(\psi) := \sup_{0<s<\ell} \left\{ \frac{\psi(s) - b(s) + \int_0^s \frac{c(r)a'(r)}{\psi(r)}\,dr}{s} \right\} \tag{8.3}$$

and

$$\mathcal{G}_i(\psi) := \inf_{0<s<\ell} \left\{ \frac{-\psi(s) - b(\ell) + b(s) + \int_s^\ell \frac{c(r)a'(r)}{\psi(r)}\,dr}{\ell - s} \right\}$$

on $\mathcal{R}$. Set $\sigma_i := \inf\{\mathcal{F}_s(\psi) : \psi \in \mathcal{S}_0\}$ and $\sigma_s := \sup\{\mathcal{F}_i(\psi) : \psi \in \mathcal{S}_1\}$. Then if $\sigma_i < \sigma_s$ equation (1.1) has a wavefront solution from ℓ to 0 for all wave speeds $\sigma_i < \sigma < \sigma_s$, such a solution with wave speed σ_i if and only if $\sigma_i = \mathcal{F}_s(\psi)$ for some $\psi \in \mathcal{S}_0$, and, such a solution with wave speed σ_s if and only if $\sigma_s = \mathcal{F}_i(\psi)$ for some $\psi \in \mathcal{S}_1$. If $\sigma_i = \sigma_s$ equation (1.1) has a wavefront solution from ℓ to 0 with wave speed σ_i if and only if $\sigma_i = \mathcal{F}_s(\psi)$ for some $\psi \in \mathcal{S}_0$ and $\sigma_s = \mathcal{F}_i(\psi)$ for some $\psi \in \mathcal{S}_1$. In any event, equation (1.1) has no wavefront solution from ℓ to 0 for any wave speed $\sigma < \sigma_i$ nor any wave speed $\sigma > \sigma_s$. Furthermore, when $c(u) \geq 0$ for all $0 < u < \ell$, there holds $\sigma_i = \inf\{\mathcal{G}_s(\psi) : \psi \in \mathcal{R}\}$, and, $\sigma_i = \mathcal{F}_s(\psi)$ for some $\psi \in \mathcal{S}_0$ if and only if $\sigma_i = \mathcal{G}_s(\psi)$ for some $\psi \in \mathcal{R}$. Whereas, when $c(u) \leq 0$ for all $0 < u < \ell$, there holds $\sigma_s = \sup\{\mathcal{G}_i(\psi) : \psi \in \mathcal{R}\}$, and, $\sigma_s = \mathcal{F}_i(\psi)$ for some $\psi \in \mathcal{S}_1$ if and only if $\sigma_s = \mathcal{G}_i(\psi)$ for some $\psi \in \mathcal{R}$.

This theorem follows from Lemma 3.5 when a similar argument to that applied to prove Theorem 4.3 is used to characterize the set of values σ for which (1.9) has a solution satisfying the integrability condition on $[0,\ell]$ and the set of values σ for which (8.1) has a solution satisfying the integrability condition on $[0,\ell]$.

Complementary to Theorems 8.1 and 8.2, the next two theorems provide more specific information on the set of wave speeds for which equation (1.1) has a wavefront solution from ℓ to 0

Theorem 8.3. *Suppose that $\ell < \infty$.*

(i) *If $c(u) < 0$ for all $0 < u < \ell$, the set of wave speeds for which* (1.1) *has a wavefront solution from ℓ to 0 is either empty or a closed interval which is unbounded below and bounded above.*

(ii) *If $c(u) > 0$ for all $0 < u < \ell$, the set of wave speeds for which* (1.1) *has a wavefront solution from ℓ to 0 is either empty or a closed interval which is bounded below and unbounded above.*

(iii) *If ca' is differentiable in $(0,\ell)$, $c(u) < 0$ for all $0 < u < \delta$ and $c(u) < 0$ for all $\ell - \delta < u < \ell$ for some $0 < \delta < \ell/2$, the set of wave speeds for which* (1.1) *has a wavefront solution from ℓ to 0 is either as in part* (i) *or a bounded interval which contains its right endpoint but not its left.*

(iv) *If ca' is differentiable in $(0,\ell)$, $c(u) > 0$ for all $0 < u < \delta$ and $c(u) > 0$ for all $\ell - \delta < u < \ell$ for some $0 < \delta < \ell/2$, the set of wave speeds for which* (1.1) *has a wavefront solution from ℓ to 0 is either as in part* (ii) *or a bounded interval which contains its left endpoint but not its right.*

(v) *If $c(u) \leq 0$ for all $0 < u < \ell$, or, if ca' is differentiable in $(0,\ell)$ and $c(u) \leq 0$ for all $0 < u < \delta$ for some $0 < \delta < \ell$, the set of wave speeds for which* (1.1) *has a wavefront solution from ℓ to 0 is either empty, contains a single value, or is an interval which is bounded above and contains its right endpoint.*

(vi) *If $c(u) \geq 0$ for all $0 < u < \ell$, or, if ca' is differentiable in $(0,\ell)$ and $c(u) \geq 0$ for all $\ell - \delta < u < \ell$ for some $0 < \delta < \ell$, the set of wave speeds for which* (1.1) *has a wavefront solution from ℓ to 0 is either empty, contains a single value, or is an interval which is bounded below and contains its left endpoint.*

(vii) *If ca' is differentiable in $(0,\ell)$, $c(u) \leq 0$ for all $0 < u < \delta$ and $c(u) \geq 0$ for all $\ell - \delta < u < \ell$ for some $0 < \delta < \ell/2$, the set of wave speeds for which* (1.1) *has a wavefront solution from ℓ to 0 is either empty or contains a single value.*

Theorem 8.4. *Consider equation* (1.1) *with two different sets of coefficients a_i, b_i, c_i on some bounded interval $[0,\ell]$ for $i = 1,2$.*

(a) *Suppose that $b_2(u) - b_2(\ell) \geq b_1(u) - b_1(\ell)$ and $(c_1a_1')(u) \leq (c_2a_2')(u) < 0$ for all $0 < u < \ell$.*

(b) *Suppose that $b_2(u) \geq b_1(u)$ and $(c_1a_1')(u) \geq (c_2a_2')(u) > 0$ for all $0 < u < \ell$.*

Then in both cases (a) *and* (b) *the set of wave speeds for which* (1.1) *with* $i = 1$ *admits a wavefront solution from* ℓ *to* 0 *is a subset of the wave speeds for which* (1.1) *with* $i = 2$ *admits a solution of this type.*

The proof of these theorems is aided by the next two lemmata.

Lemma 8.5. *Fix* $0 < \delta \le \ell$ *with* $\delta < \infty$, *and let* S *denote the set of values* σ *for which equation* (1.9) *has a solution satisfying the integrability condition on* $[0, \delta]$. *Suppose that* S *is not empty, define* $\sigma_i := \inf S$, *and let* $\theta(\cdot;\sigma)$ *denote the maximal solution of* (1.9) *on* $[0, \delta]$ *for all* $\sigma \in S$. *Then,*

$$S = \begin{cases} [\sigma_i, \infty) & \text{if } \sigma_i \in S \\ (\sigma_i, \infty) & \text{if } \sigma_i \notin S, \end{cases}$$

the variable $\overline{\theta}(s;\sigma^*) := \lim_{\sigma \downarrow \sigma^*} \theta(s;\sigma)$ *is well-defined for all* $0 \le s \le \delta$ *and* $\sigma^* \in S$, *and, the variable* $\underline{\theta}(s;\sigma^*) := \lim_{\sigma \uparrow \sigma^*} \theta(s;\sigma)$ *is well-defined for all* $0 \le s \le \delta$ *and* $\sigma^* > \sigma_i$. *Next, for fixed* $\sigma \in S$, *let* R *denote the set of values* $\theta(\delta)$ *such that equation* (1.9) *has a solution* θ *satisfying the integrability condition on* $[0, \delta]$, *and define* $\rho_i = \inf R$. *Then,*

$$R = \begin{cases} [\rho_i, \overline{\theta}(\delta;\sigma)] & \text{if } \rho_i \in R \\ (\rho_i, \overline{\theta}(\delta;\sigma)] & \text{if } \rho_i \notin R, \end{cases}$$

$\underline{\theta}(\delta;\sigma) \in R$ *if* $\sigma > \sigma_i$, *and, given any* $\rho \in R$ *there exists a unique maximal solution* $\theta(\cdot;\sigma;\rho)$ *of* (1.9) *on* $[0, \delta]$ *taking the value* ρ *in* δ. *There holds* $\theta(s;\sigma;\overline{\theta}(\delta;\sigma)) = \theta(s;\sigma)$ *for all* $0 \le s \le \delta$, *and,* $\theta(s;\sigma;\rho^{(1)}) \le \theta(s;\sigma;\rho^{(2)})$ *for all* $0 \le s \le \delta$ *and* $\rho^{(1)}, \rho^{(2)} \in R$ *with* $\rho^{(1)} \le \rho^{(2)}$.

Proof. Lemmata 2.18 and 2.17 tell us that S is connected and unbounded above and that $\theta(s;\sigma)$ is a monotonic function of $\sigma \in S$ for all $0 \le s \le \delta$. Thus the assertions concerning S and the definitions of $\underline{\theta}$ and $\overline{\theta}$ are proven. A straightforward limit argument (cf. [123]) then shows that $\underline{\theta}(\cdot;\sigma)$ is a solution of (1.9) on $[0, \delta]$ for every $\sigma > \sigma_i$, and that $\overline{\theta}(\cdot;\sigma)$ is a solution of (1.9) on $[0, \delta]$ for every $\sigma \in S$. The uniqueness of the maximal solution hereafter implies that $\underline{\theta}(s;\sigma) \le \theta(s;\sigma)$ for all $0 \le s \le \delta$ and $\sigma > \sigma_i$, and, $\overline{\theta}(s;\sigma) = \theta(s;\sigma)$ for all $0 \le s \le \delta$ and $\sigma \in S$. Now, suppose that for some $\sigma \in S$ equation (1.9) admits a solution θ^* satisfying the integrability condition on $[0, \delta]$ with $\theta^*(\delta) \le \theta(\delta;\sigma)$. Then by Lemmata 2.13, 2.17 and 2.18, for any $\theta^*(\delta) \le \rho \le \theta(\delta;\sigma)$, the equation

$$\theta(s) = \rho + \sigma(s - \delta) + b(s) - b(\delta) + \int_s^\delta \frac{c(r)a'(r)}{\theta(r)}\,dr \tag{8.4}$$

has a solution θ on $[0, \delta]$ such that $\theta^*(s) \le \theta(s) \le \theta(s;\sigma)$ for all $s \in [0, \delta]$. Furthermore, it is possible to construct a unique maximal solution $\theta(\cdot;\sigma;\rho)$ of (8.4) with this property. Subsequently, it can be seen that $\theta(\cdot;\sigma;\rho)$ is a solution of (1.9) satisfying the integrability condition on $[0, \ell]$ and $\theta(\delta;\sigma;\rho) = \rho$. This proves the assertions concerning R. The assertions concerning $\{\theta(\cdot;\sigma;\rho) : \rho \in R\}$ follow from the general theory of equations of the form (8.4) in Section 2.2. □

Lemma 8.6. *Let $0 < \ell < \infty$. Suppose that $c(u) \leq 0$ for all $0 < u < \ell$ or that ca' is differentiable in $(0, \ell)$ and $c(u) \leq 0$ for all $0 < u < \delta$ for some $0 < \delta < \ell$. Suppose furthermore that the set S of values σ for which equation (1.9) has a solution θ satisfying the integrability condition on $[0, \ell]$ and $\theta(\ell) = 0$ is not empty. Then $\sigma_s := \sup S \in S$.*

Proof. Pick $\sigma_0 \in S$. Then, by definition and Lemmata 2.14(i), 2.17, 2.18, 2.40 and 4.7, equation (1.9) has a unique solution $\theta(\cdot;\sigma)$ satisfying the integrability condition on $[0, \ell]$ for every $\sigma \geq \sigma_0$. Moreover, by the previous lemma, $\theta(\ell; \sigma)$ is a continuous nondecreasing function of $\sigma \geq \sigma_0$. It follows that $\sigma_s = \sup\{\sigma \geq \sigma_0 : \theta(\ell;\sigma) = 0\}$, and, either $\sigma_s = \infty$ or $\sigma_s \in S$. However, by a lemma in [124], $\theta(\ell;\sigma) \to \infty$ as $\sigma \to \infty$. So the first alternative is excluded. □

Proof of Theorem 8.3. We shall prove parts (i), (iii), (v) and (vii) only, since the remaining parts may be obtained from these via Theorem 3.4.

(i) The key to the first part of the theorem is the observation that if $c < 0$ on $(0, \ell)$, equation (1.9) automatically admits a solution on $[0, \ell]$ which is positive on $(0, \ell)$ by Lemmata 2.16(i) and 2.17; while, any solution of equation (8.1) on $[0, \ell]$ automatically satisfies the integrability condition on $[0, \ell]$ by Lemma 2.24. The task of finding wave speeds for which (1.1) has a wavefront solution from ℓ to 0 is therefore reduced to the task of finding σ for which (8.1) has a solution on $[0, \ell]$. In this light, the result follows from Lemmata 8.5 and 8.6.

(iii) Let S denote that set of numbers σ such that equation (1.9) has a solution θ on $[0, \ell]$ which is positive on $(0, \ell)$ and satisfies $\theta(\ell) = 0$. Then, in view of Theorem 2.10, Lemmata 2.40 and 8.6, to prove this part of the theorem, it suffices to show that $\sigma_i := \inf S \notin S$. To achieve this we observe to begin with that by Lemma 2.16(i), equation (1.9) has a unique solution $\theta(\cdot;\sigma)$ in a right neighbourhood of zero for every σ. We let $\Delta(\sigma)$ be such that $\theta(\cdot;\sigma)$ is defined on $[0, \Delta(\sigma)]$, positive on $(0, \Delta(\sigma))$, and, $\Delta(\sigma) = \ell$ or $\theta(\Delta(\sigma);\sigma) = 0$. Lemmata 2.16(i), 2.17, 2.18, 2.40 and 2.24, imply that $\sigma \mapsto \Delta(\sigma)$ is a well-defined nondecreasing function $(-\infty, \infty) \to (0, \ell]$. We assert that this function is also continuous from the left. To verify this assertion, suppose to its contrary that there exist a σ^* such that $\Delta^* := \sup\{\Delta(\sigma) : \sigma < \sigma^*\} < \Delta(\sigma^*)$. Pick $0 < \rho < \theta(\Delta^*;\sigma^*)$. By Lemmata 2.13 and 2.17, for every σ there exists a $0 \leq s_0(\sigma) < \Delta^* < s_1(\sigma) \leq \ell$ such that equation (8.4) has a unique solution $\theta^*(\cdot;\sigma)$ on $[s_0(\sigma), s_1(\sigma)]$, which is positive on $(s_0(\sigma), s_1(\sigma))$, and, such that $s_0(\sigma) = 0$ or $\theta^*(s_0(\sigma);\sigma) = 0$. In particular, since $\theta^*(s;\sigma^*) < \theta(s;\sigma^*)$ for all $s_0(\sigma^*) < s \leq \Delta^*$ by Lemmata 2.18 and 4.7, there holds $0 < s_0(\sigma^*) < \Delta^*$ and $\theta^*(s_0(\sigma^*);\sigma^*) = 0$. Furthermore, Lemmata 2.18 and 4.7 imply that s_0 and s_1 are nondecreasing functions of σ, and, for fixed $0 < s < \ell$, that $\theta(s;\sigma)$ is a nondecreasing function of $\sigma > \sigma_i$ whenever $\Delta(\sigma) > s$. Consequently, because $\theta^*(s_0(\sigma^*);\sigma) \to \theta^*(s_0(\sigma^*);\sigma^*) = 0$ as $\sigma \uparrow \sigma^*$ by extension of the argument in Lemma 8.5, we can find a $\sigma^{**} < \sigma^*$ so large that $\Delta(\sigma) > s_0(\sigma^*)$ and

$\theta(s_0(\sigma^*);\sigma) > \theta^*(s_0(\sigma^*);\sigma)$ for all $\sigma > \sigma^{**}$. However, by Lemmata 2.18 and 4.7, this means that $\Delta(\sigma) \geq s_1(\sigma) \geq s_1(\sigma^{**}) > \Delta^*$ for all $\sigma^{**} < \sigma < \sigma^*$. This contradicts our original supposition. So we conclude that $\sigma \mapsto \Delta(\sigma)$ is indeed continuous from the left. Now suppose that $\sigma_i \in S$. Then, $\Delta(\sigma_i) = \ell > \Delta(\sigma)$ for any $\sigma < \sigma_i$; while, $\Delta(\sigma) \to \Delta(\sigma_i)$ as $\sigma \uparrow \sigma_i$. Hence there exists a $\sigma^* < \sigma_i$ such that $\Delta(\sigma^*) > \ell - \delta$. For every $\sigma^* < \sigma < \sigma_i$ this implies that $\Theta(s) := \theta(\Delta(\sigma) - s;\sigma)$ is a solution of equation (8.1) with ℓ replaced by $\Delta(\sigma)$ on $[0, \Delta(\sigma)]$, while, by Lemma 2.15(ii) such a solution cannot exist. Thus, the supposition $\sigma_i \in S$ is refuted.

(v) This part of the theorem is a straightforward corollary of Theorem 2.10 and Lemma 8.6.

(vii) To verify this part of the theorem, suppose that there are two wave speeds σ_1 and $\sigma_2 \geq \sigma_1$ such that equation (1.1) has a wavefront solution from ℓ to 0. Then by Theorem 2.10 and Lemma 2.40 there exists a solution θ_i of (1.9) with $\sigma = \sigma_i$ on $[0, \ell]$ which is positive on $(0, \ell)$ and such that $\theta_i(\ell) = 0$ for $i = 1, 2$. Subsequently, the function Θ_i, defined by $\Theta_i(s) := \theta_i(\ell - s)$ for $0 \leq s \leq \ell$, is a solution of (8.1) on $[0, \ell]$ with similar properties. By Lemmata 2.14(i), 2.18(i) and 4.7 there holds $\theta_2(s) \geq \theta_1(s)$ for all $s \in (0, \ell)$. On the other hand, by the same lemmata applied to equation (8.1) there holds $\Theta_2(s) \leq \Theta_1(s)$ for all $s \in (0, \ell)$. This is clearly incompatible unless $\sigma_1 = \sigma_2$. □

Proof of Theorem 8.4. From the proof of Theorem 8.3 part (i), when $c < 0$ on $(0, \ell)$, equation (1.1) admits a wavefront solution from ℓ to 0 with wave speed σ if and only if (8.1) has a solution on $[0, \ell]$. The conclusion in case (a) subsequently follows from Lemma 2.18. The conclusion in case (b) follows hereafter via Theorem 3.4. □

8.2 Number of wavefronts

Having analysed for which wave speeds an equation of the class (1.1) can admit a wavefront solution, one may enquire as to how many distinct wavefronts with any given wave speed there can be. The next theorem provides a partial answer to this enquiry.

Theorem 8.7. *Suppose that $\ell < \infty$. Let σ be any fixed wave speed. Then equation (1.1) has at most one distinct wavefront solution from ℓ to 0 with this wave speed whenever any one of the following hold.*

(a) *$c(u) < 0$ for all $0 < u < \ell$.*

(b) *$c(u) > 0$ for all $0 < u < \ell$.*

(c) *ca' is differentiable in $(0, \ell)$ and $c(u) \leq 0$ for all $0 < u < \delta$ for some $0 < \delta \leq \ell$.*

(d) *ca' is differentiable in $(0, \ell)$ and $c(u) \geq 0$ for all $\ell - \delta < u < \ell$ for some $0 < \delta \leq \ell$.*

Proof. Parts (a) and (c) of the theorem are a simple consequence of Theorem 4.4. Parts (b) and (d) follow from (a) and (c) respectively, by application of Theorem 3.4. □

8.3 Illustrations

Up to now the discussion of wavefront solutions has been concerned with the general case. Let us now examine the consequences for two particular examples investigated previously by other authors. Because it is helpful to these examples, we first introduce two lemmata. The second represents a variation on Lemma 3.5.

Lemma 8.8. *Suppose that $c(s) \geq 0$ for all $0 < s < \delta$ for some $0 < \delta < \ell$. Define*

$$\beta := \limsup_{s \downarrow 0} \frac{b(s)}{s}$$

and λ_0 by (6.15). *Then if* (1.9) *has a solution θ, necessarily $\sigma \geq 2\sqrt{\lambda_0} - \beta$ and*

$$\limsup_{s \downarrow 0} \frac{\theta(s)}{s} \leq \frac{\sigma + \beta + \sqrt{(\sigma + \beta)^2 - 4\lambda_0}}{2}.$$

Proof. By a direct copy of the proof of part (i) of Lemma 6.6 it can be shown that (1.9) has a solution θ on $[0, \delta)$ only if $\sigma \geq 2\sqrt{\Lambda_0} - B$ and $\theta(s) \leq \mu s$ for all $0 < s < \delta$, where Λ_0 is given by (6.18), $B := \sup\{b(s)/s : 0 < s < \delta\}$, and $\mu := \{\sigma + B + \sqrt{(\sigma + B)^2 - 4\Lambda_0}\}/2$. The assertion follows by noting that in this copy, δ may be chosen arbitrarily small. □

Lemma 8.9. *Suppose that $0 < \alpha < \ell < \infty$. Then equation* (1.9) *has a solution on $[0, \ell]$ taking the value 0 in ℓ, only if* (1.9) *has a solution θ on $[0, \alpha]$ and* (8.1) *has a solution Θ on $[0, \ell - \alpha]$ such that $\theta(\alpha) = \Theta(\ell - \alpha)$. Conversely, equation* (1.9) *has a solution on $[0, \ell]$ taking the value 0 in ℓ, if* (1.9) *has a solution θ on $[0, \alpha]$ and* (8.1) *has a solution Θ on $[0, \ell - \alpha]$ such that $\theta(\alpha) = \Theta(\ell - \alpha) > 0$. Idem ditto, equation* (1.9) *has a solution satisfying the integrability condition on $[0, \ell]$ and taking the value 0 in ℓ, only if* (1.9) *has a solution θ satisfying the integrability condition on $[0, \alpha]$ and* (8.1) *has a solution Θ satisfying the integrability condition on $[0, \ell - \alpha]$ such that $\theta(\alpha) = \Theta(\ell - \alpha)$. Conversely, equation* (1.9) *has a solution satisfying the integrability condition on $[0, \ell]$ and taking the value 0 in ℓ, if* (1.9) *has a solution θ satisfying the integrability condition on $[0, \alpha]$ and* (8.1) *has a solution Θ satisfying the integrability condition on $[0, \ell - \alpha]$ such that $\theta(\alpha) = \Theta(\ell - \alpha) > 0$.*

Proof. The 'only if' assertions of the lemma have been previously established in the course of the proof of Lemma 3.5. Furthermore, under the necessary conditions it can be checked that the function ψ defined by

$$\psi(s) := \begin{cases} \theta(s) & \text{for } 0 \leq s \leq \alpha \\ \Theta(\ell - s) & \text{for } \alpha < s \leq \ell, \end{cases}$$

is a candidate for a solution of (1.9) on $[0, \ell]$ taking the value 0 in ℓ. The only disputable point is whether ca'/ψ has sufficient integrability properties in a neighbourhood of α to qualify ψ as a solution of (1.9). The condition $\psi(\alpha) > 0$ takes care of this. Likewise, if θ and Θ satisfy the integrability condition on $[0, \alpha]$ and $[0, \ell - \alpha]$ respectively, the condition $\psi(\alpha) > 0$ ensures that ψ satisfies the integrability condition on the whole of $[0, \ell]$. □

Our first example is a classic one from mathematical biology, showing the influence of including a nonlinear convection term in the Fisher equation.

Application 8.10. *The equation*

$$u_t + kuu_x = u_{xx} + u(1-u), \tag{8.5}$$

where k is a real constant, admits a wavefront solution from 1 to 0 with wave speed σ if and only if $\sigma \geq \sigma^*$, where

$$\sigma^* := \begin{cases} 2 & \text{if } k \leq 2 \\ (k^2+4)/2k & \text{if } k > 2. \end{cases}$$

Furthermore, for each wave speed $\sigma \geq \sigma^*$ the wavefront solution f is unique modulo translation, and such that $0 < f(\xi) < 1$ and

$$-A \geq \frac{f'(\xi)}{f(\xi)\{1 - f(\xi)\}} \geq -B \quad \text{for all } -\infty < \xi < \infty$$

where A and B are as stated below.

(a) *If $k \leq 0$, or, if $0 < k < 2$ and $\sigma < (k^2+4)/2k$,*

$$A = \frac{2}{\sigma - k + \sqrt{(\sigma-k)^2+4}} \quad \text{and} \quad B = \frac{2}{\sigma + \sqrt{\sigma^2 - 4}}.$$

(b) *If $k > 0$ and $\sigma = (k^2+4)/2k$,*

$$A = B = \frac{k}{2}.$$

(c) *If $k > 0$ and $\sigma > (k^2+4)/2k$,*

$$A = \frac{2}{\sigma + \sqrt{\sigma^2-4}} \quad \text{and} \quad B = \frac{2}{\sigma - k + \sqrt{(\sigma-k)^2+4}}.$$

Proof. Equation (8.5) is of the form (1.1) with $a(u) = u, b(u) = -ku^2/2$ and $c(u) = u(1-u)$. Subsequently, by Lemma 3.5, the equation admits a wavefront solution of the sought-after type if and only if the integral equations

$$\theta(s) = \sigma s - \frac{k}{2}s^2 - \int_0^s \frac{r(1-r)}{\theta(r)}\,dr \tag{8.6}$$

and

$$\Theta(s) = -\sigma s + \frac{k}{2}s(2-s) + \int_0^s \frac{r(1-r)}{\Theta(r)}\,dr \tag{8.7}$$

admit solutions satisfying the integrability condition on $[0,1]$. However, by Lemmata 2.16(i) and 2.17, equation (8.7) admits a unique solution on $[0,1]$ which is positive on $(0,1)$ for any σ. Whereas, by Lemma 8.8, equation (8.6) admits a solution θ on $[0,1]$ only if $\sigma \geq 2$ and

$$\limsup_{s\downarrow 0} \frac{\theta(s)}{s} \leq \frac{\sigma + \sqrt{\sigma^2-4}}{2}. \tag{8.8}$$

To proceed we distinguish between the cases $k \leq 2$ and $k > 2$. For $k \leq 2$ we observe that the function $\theta(s) := s(1-s)$ satisfies (8.6) with $\sigma = 2$ and $k = 2$ on $[0,\ell]$. Thus, by Lemma 2.18, equation (8.6) has a solution for any $\sigma \geq 2$ and $k \leq 2$ on $[0,\ell]$. On the other hand, for $k > 2$ we observe that $\theta_1(s) := ks(1-s)/2$ satisfies (8.6) with $\sigma = (k^2+4)/2k$ on $[0,\ell]$. Hence, by Lemma 2.18, equation (8.6) has a solution for any $\sigma \geq (k^2+4)/2k$ and $k > 2$ on $[0,\ell]$. Furthermore, should θ denote a solution of (8.6) with $2 \leq \sigma < (k^2+4)/2k$ on $[0,\ell]$, then setting $\Theta(s) := \theta(1-s)$ and $\Theta_1(s) := \theta_1(1-s)$ by Lemmata 2.14(i) and 2.18(i) applied to (8.7) necessarily $\Theta(s) > \Theta_1(s)$ for all $0 < s < 1$, and hence

$$\limsup_{s\downarrow 0} \frac{\theta(s)}{s} \geq \limsup_{s\downarrow 0} \frac{\theta_1(s)}{s} = \frac{k}{2}. \tag{8.9}$$

Combining (8.8) and (8.9) yields $\sigma + \sqrt{\sigma^2-4} \geq k$, which after some elementary manipulation can be shown to contradict the assumption that $2 \leq \sigma < (k^2+4)/2k$ and $k > 2$. Thus for this range of parameter values, such a solution θ cannot exist. This confirms the existence result for the wavefront solutions. The uniqueness result is given by Theorem 8.7. The remaining assertions of the example follow from estimates of the unique solution of (8.7). We note that the function $\Theta_1(s) = As(1-s)$ is a solution of (8.7) on $[0,1]$ with $-\sigma s + ks(2-s)/2$ replaced by $b_1(s) := s(A - 1/A - As)$. Subsequently since $s \mapsto -\sigma s + ks(2-s)/2 - b_1(s)$ is nondecreasing on $[0,1]$, there holds $\Theta(s) \geq \Theta_1(s)$ for the unique solution Θ of (8.7) on $[0,1]$. Similarly we deduce that $\Theta(s) \leq \Theta_2(s) := Bs(1-s)$ for all $0 \leq s \leq 1$. This gives the required result noting that for this particular example the correspondence between a wavefront solution f of (1.1) and a solution θ of the integral equation is given by $f' = -\theta(f) = -\Theta(1-f)$. □

The second example we consider concerns a simplified model for reacting gas flows. It extends known results on reaction-convection-diffusion equations generalizing the Nagumo equation.

Application 8.11. *Suppose that ca' is differentiable in $(0,\ell)$, $c(u) < 0$ for all $0 < u < \alpha$, $c(u) > 0$ for all $\alpha < u < \ell$, and, $(ca')'(\alpha) > 0$, for some $0 < \alpha < \ell < \infty$. Then there exists a unique wave speed σ such that equation (1.1) has a wavefront solution from ℓ to 0, and, this wavefront solution is unique modulo translation.*

Proof. By Lemmata 2.16(i) and 2.17 equation (1.9) has a unique solution $\theta(\cdot;\sigma)$ on $[0,\alpha]$ which is positive on $(0,\alpha)$ and equation (8.1) has a unique solution $\Theta(\cdot;\sigma)$ on $[0,\ell-\alpha]$ which is positive on $(0,\ell-\alpha)$ for any σ. Furthermore, by Lemma 8.5, the function $F(\sigma) := \theta(\alpha;\sigma) - \Theta(\ell-\alpha;\sigma)$ depends continuously on σ, and, by a lemma in [124], $F(\sigma) \to \pm\infty$ as $\sigma \to \pm\infty$. Consequently there exists at least one value σ such that $F(\sigma) = 0$. We assert that for such a value necessarily $\theta(\alpha;\sigma) = \Theta(\ell-\alpha;\sigma) > 0$. For if this is not the case, $\widetilde{\Theta}(s) := \theta(\alpha-s;\sigma)$ defines a solution of

$$\widetilde{\Theta}(s) = -\sigma s + b(\alpha - s) - b(\alpha) + \int_0^s \frac{c(\alpha-r)a'(\alpha-r)}{\widetilde{\Theta}(r)}\,dr$$

on $[0,\alpha]$, and $\widetilde{\theta}(s) := \Theta(\ell-\alpha-s)$ defines a solution of

$$\widetilde{\theta}(s) = \sigma s + b(\alpha + s) - b(\alpha) - \int_0^s \frac{c(\alpha+r)a'(\alpha+r)}{\widetilde{\theta}(r)}\,dr$$

on $[0,\ell-\alpha]$. However, by Lemma 8.8 for the existence of $\widetilde{\Theta}$, necessarily

$$-\sigma \geq 2\sqrt{(ca')'(\alpha)} + b'(\alpha), \tag{8.10}$$

and for the existence of $\widetilde{\theta}$, necessarily

$$\sigma \geq 2\sqrt{(ca')'(\alpha)} - b'(\alpha), \tag{8.11}$$

while, (8.10) and (8.11) are incompatible because $(ca')'(\alpha) > 0$. Thus there exists at least one real value σ such that $\theta(\alpha;\sigma) = \Theta(\ell-\alpha;\sigma) > 0$. Subsequently by Lemma 8.9 and Theorem 2.10, equation (1.1) has a wavefront solution from ℓ to 0. The uniqueness of the wave speed and of the wavefront are provided by Theorems 8.3(vii) and 8.7 respectively. □

8.4 Multiple equilibria

When $c(0) = c(\alpha) = c(\ell) = 0$ for some $0 < \alpha < \ell < \infty$, it is conceivable that equations of the class (1.1) admit wavefront solutions from ℓ to 0, from ℓ to α, and, from α to 0. The next two theorems are concerned with the relationship between the corresponding sets of admissible wave speeds.

Theorem 8.12. *Suppose that* $c(\alpha) = 0$ *for some* $0 < \alpha < \ell < \infty$, *and equation* (1.1) *has a wavefront solution from* ℓ *to* 0 *with wave speed* σ^*.

(i) *Suppose furthermore that* (1.1) *has a wavefront solution from* α *to* 0 *with wave speed* σ_0. *Then, either* $\sigma_0 < \sigma^*$, *or,* (1.1) *has a wavefront solution from* α *to* 0 *with wave speed* σ *for every* $\sigma^* \leq \sigma \leq \sigma_0$. *In particular, if* $c(u) \leq 0$ *for all* $0 < u < \alpha$, *or,* ca' *is differentiable in* $(0,\alpha)$ *and* $c(u) \leq 0$ *for all* $0 < u < \delta$ *for some* $0 < \delta < \alpha$, *and, if* (1.1) *has exactly one distinct wavefront solution from* ℓ *to* 0 *with wave speed* σ^*, *then* $\sigma_0 < \sigma^*$.

(ii) *Suppose furthermore that* (1.1) *has a wavefront solution from ℓ to α with wave speed σ_1. Then, either $\sigma_1 > \sigma^*$, or,* (1.1) *has a wavefront solution from ℓ to α with wave speed σ for every $\sigma_1 \leq \sigma \leq \sigma^*$. In particular, if $c(u) \geq 0$ for all $\alpha < u < \ell$, or, ca' is differentiable in (α, ℓ) and $c(u) \geq 0$ for all $\ell - \delta < u < \delta$ for some $0 < \delta < \ell - \alpha$, and, if* (1.1) *has exactly one distinct wavefront solution from ℓ to 0 with wave speed σ^*, then $\sigma_1 > \sigma^*$.*

Proof. We prove part (i) only, since part (ii) may be proven similarly or obtained from part (i) via Theorem 3.4. By Theorem 2.10, equation (1.9) with $\sigma = \sigma^*$ has a solution satisfying the integrability condition on $[0, \ell]$, and, thus also on $[0, \alpha]$. Simultaneously, by Theorem 2.10 and Lemma 3.5, equation (8.1) with $\sigma = \sigma_0$ and ℓ replaced by α has a solution satisfying the integrability condition on $[0, \alpha]$. Lemma 2.18 then implies that (1.9) has a solution satisfying the integrability condition on $[0, \alpha]$ for every $\sigma \geq \sigma^*$, while (8.1) with ℓ replaced by α has a solution satisfying the integrability condition on $[0, \alpha]$ for every $\sigma \leq \sigma_0$. The main assertion of part (i) of the theorem subsequently follows from Theorem 2.10 and Lemma 3.5. Suppose now that $\sigma_0 \geq \sigma^*$. Then, by what we have already established, equation (1.9) with $\sigma = \sigma^*$ has a solution θ_1 satisfying the integrability condition on $[0, \ell]$ such that $\theta_1(\ell) = 0$, and, a solution θ_2 satisfying the integrability condition on $[0, \alpha]$ such that $\theta_2(\alpha) = 0$. However, if $c(u) \leq 0$ for all $0 < u < \alpha$, or, ca' is differentiable in $(0, \alpha)$ and $c(u) \leq 0$ for all $0 < u < \delta$ for some $0 < \delta < \alpha$, Lemmata 2.14(i), 2.17, 2.40 and 4.7 tell us that $\theta_1 \equiv \theta_2$ on $[0, \alpha]$. Hence, in particular, $\theta_1(\alpha) = 0$. Lemma 2.26 subsequently infers that (1.1) has more than one distinct wavefront solution from ℓ to 0 with wave speed σ^*. Thus, the subsidiary assertion of part (i) of the theorem is established by a contradiction argument. □

Theorem 8.13. *Suppose that $c(\alpha) = 0$ for some $0 < \alpha < \ell < \infty$. Suppose furthermore that equation* (1.1) *has a wavefront solution from α to 0 with wave speed σ_0, and, a wavefront solution from ℓ to α with wave speed σ_1. Then if $\sigma_0 < \sigma_1$, and, there is no $\sigma_0 \leq \sigma \leq \sigma_1$ such that* (1.1) *has a both a wavefront solution from α to 0 and a wavefront solution from ℓ to α with wave speed σ, there exists a $\sigma_0 \leq \sigma^* \leq \sigma_1$ such that* (1.1) *has a wavefront solution from ℓ to 0 with wave speed σ^*.*

Proof. The proof of this theorem is a refinement of the argument used to confirm Application 8.11. Since equation (1.1) has a wavefront solution from α to 0 with wave speed σ_0, equation (1.9) with $\sigma = \sigma_0$ has a solution θ_0 satisfying the integrability condition on $[0, \alpha]$ and $\theta_0(\alpha) = 0$. Subsequently, adopting the notation of Lemma 8.5, we can define the set $\Gamma_0 := \{(\sigma_0, \rho) : 0 \leq \rho \leq \theta(\alpha; \sigma_0)\} \cup \{(\sigma, \rho) : \underline{\theta}(\alpha; \sigma) \leq \rho \leq \theta(\alpha; \sigma)$ and $\sigma > \sigma_0\}$. By Lemma 8.5, Γ_0 is a continuous monotonic graph, and, for every $(\sigma, \rho) \in \Gamma_0$ equation (1.9) has a unique maximal solution θ satisfying the integrability condition on $[0, \alpha]$ and $\theta(\alpha) = \rho$. Similarly, equation (8.1) with $\sigma = \sigma_1$ admits a solution Θ_1 satisfying the integrability condition on $[0, \ell - \alpha]$ and $\Theta_1(\ell - \alpha) = 0$. Furthermore, we can define a continuous monotone graph Γ_1 with the properties $(\sigma_1, 0) \in \Gamma_1$, $\{\sigma : (\sigma, \rho) \in \Gamma_1\} = (-\infty, \sigma_1]$, and, for every $(\sigma, \rho) \in \Gamma_1$ there exists a unique maximal solution Θ of equation (8.1) satisfying the integrability condition on $[0, \ell - \alpha]$ and $\Theta(\ell - \alpha) = \rho$. Now, by the continuity of Γ_0 and Γ_1, there exists a $(\sigma^*, \rho^*) \in \Gamma_0 \cap \Gamma_1$. Moreover $\rho^* > 0$, for otherwise by Theorem 2.10 equation (1.1) would have both a wavefront solution from α to 0 and a wavefront solution from ℓ to α with wave speed σ^*. This gives the result via Lemma 8.9 and Theorem 2.10. □

Bibliographic notes

Interest in wavefront solutions can be traced through [25,56,98,101,186,211,296, 297].

Theorem 8.2 generalizes the 'variational principle' for wavefront solutions of equation (1.1) propounded by Hadeler and Rothe [150] and by Hadeler [143]. See also [82,141,142,144,145,295,297].

Theorem 8.3 part (vii) has been proven earlier for the specific case that (1.1) has no convection term and is semilinear, i.e. has the form $u_t = u_{xx} + c(u)$, by Fife and McLeod [103].

The biological background to Application 8.10 is described in the books of Murray [210,211]. Besides in [210,211], Application 8.10 has been previously examined by Kelley [179]. The estimates on f' in this example show that as the wave speed $\sigma \to \infty$, then f' behaves like $-f(1-f)/\sigma$ which conclusion was previously reached in [179] using dimensional analysis. The motivation for Application 8.11 as a model of reacting gas flows is given in [280].

For the semilinear reaction-diffusion equation $u_t = u_{xx} + c(u)$, where c has zeros in 0, α and ℓ, $0 < \alpha < \ell < \infty$, the relationship between the sets of wave speeds for which the equation may admit wavefront solutions from ℓ to 0, from ℓ to α, and, from α to 0 has been studied heuristically by Shkadinskii, Barelko and Kurochka [270]. Under the assumption that $c \in C^1([0, \ell])$, this relationship has been examined in more depth by Fife and McLeod [103], by Vol'pert [296], and, by Vol'pert et al. [297]. In [296–298] a family of such solutions is called a *system of waves*. Theorems 8.12 and 8.13 extend the findings in [103,296,297] to the general case. We shall return to the subject of Theorems 8.12 and 8.13 in the bibliographic notes of Chapter 10, when we are armed with more detailed analysis of wavefront solutions of reaction-diffusion equations.

Chapter 9

Wavefronts for convection-diffusion

As mentioned earlier, when the reaction term in (1.1) is absent and the partial differential equation is

$$u_t = (a(u))_{xx} + (b(u))_x, \tag{9.1}$$

the integral equation (1.9) reduces to the simple identity $\theta(s) = \sigma s + b(s)$. Subsequently, if this 'equation' is to have a nonnegative solution on $[0, \ell]$ with $\ell < \infty$ such that $\theta(\ell) = 0$, then necessarily $\sigma = -b(\ell)/\ell$ and $\theta(s) = b(s) - sb(\ell)/\ell \geq 0$ for all $0 \leq s \leq \ell$. By Lemma 2.40 though θ satisfies the integrability condition if and only if it is positive on $(0, \ell)$. We conclude the following.

Theorem 9.1. *Suppose that $\ell < \infty$. Then equation* (9.1) *has a wavefront solution from ℓ to 0 if and only if $\ell b(u) > ub(\ell)$ for all $0 < u < \ell$, in which case the wave speed $\sigma = -b(\ell)/\ell$. Furthermore, the solution f is unique modulo translation, and if it is so translated that $f(0) = \nu$ for some $0 < \nu < \ell$ is given by*

$$\begin{cases} f(\xi) = \ell & \text{for } \xi \leq \Xi_1 \\ \displaystyle\int_{f(\xi)}^{\nu} \frac{\ell a'(s)}{\ell b(s) - sb(\ell)}\, ds = \xi & \text{for } \Xi_1 < \xi < \Xi_0 \\ f(\xi) = 0 & \text{for } \xi \geq \Xi_0 \end{cases} \tag{9.2}$$

where

$$\Xi_0 := \int_0^{\nu} \frac{\ell a'(s)}{\ell b(s) - sb(\ell)}\, ds \quad \text{and} \quad \Xi_1 := -\int_{\nu}^{\ell} \frac{\ell a'(s)}{\ell b(s) - sb(\ell)}\, ds. \tag{9.3}$$

A corollary of this theorem is that if $\Xi_0 < \infty$, then there exists a ξ^* such that $f(\xi) = 0$ for all $\xi \geq \xi^*$, while if $\Xi_0 = \infty$, then necessarily $f(\xi) > 0$ for all ξ. Similarly, if $\Xi_1 > -\infty$, then there exists a ξ^{**} such that $f(\xi) = \ell$ for all $\xi \leq \xi^{**}$, while on the other hand if $\Xi_1 = -\infty$, then $f(\xi) < \ell$ for all ξ.

Note the similarity between the above result and that which can be obtained for the hyperbolic conservation law

$$u_t = (b(u))_x. \tag{9.4}$$

Following Kruzhkov [188, 189] and in analogy to Definition 2.2, when b satisfies Hypothesis 2.1(ii) we say that a function f defined in an open interval Ω is a travelling-wave *entropy* solution of equation (9.4) with speed σ if $f \in L^\infty_{\text{loc}}(\Omega)$, and

$$\int_\Omega [\sigma\,|f-K| + \text{signum}(f-K)\,\{b(f)-b(K)\}]\,\phi'\,d\xi \le 0 \tag{9.5}$$

for any function $\phi \in C_0^\infty(\Omega)$ and number $K \in I$. Furthermore, noting that any bounded monotonic function is continuous almost everywhere, when $\ell < \infty$ a wavefront entropy solution of equation (9.4) from ℓ to 0 may be defined as in Definition 2.4. Analysis of (9.5) then shows that for the existence of a wavefront entropy solution of equation (9.4) from ℓ to 0 it is necessary that $\ell b(u) \ge ub(\ell)$ for all $0 < u < \ell$, and that the wave speed must be $\sigma = -b(\ell)/\ell$ for this equation too. Moreover, if $\ell b(u) > ub(\ell)$ for all $0 < u < \ell$, it can be determined that the wavefront entropy solution of (9.4) is distinct, and modulo translation is given by the 'vanishingviscosity' limit of (9.2) and (9.3), i.e.

$$f(\xi) = \begin{cases} \ell & \text{for } \xi < 0 \\ \nu & \text{for } \xi = 0 \\ 0 & \text{for } \xi > 0 \end{cases}$$

for some $0 \le \nu \le \ell$.

As a simple illustration of Theorem 9.1, consider the Burgers equation

$$u_t + uu_x = u_{xx}. \tag{9.6}$$

Application 9.2. *Equation* (9.6) *admits a global travelling-wave solution f with speed σ such that $f(\xi) \to \ell^\pm$ as $\xi \to \pm\infty$ for some real numbers $\ell^- \ne \ell^+$ if and only if $\ell^- > \ell^+$ and $\sigma = (\ell^- + \ell^+)/2$, in which case*

$$f(\xi) = \frac{\ell^- + \ell^+}{2} - \frac{\ell^- - \ell^+}{2}\tanh\left\{\frac{\ell^- - \ell^+}{4}(\xi - \xi_0)\right\}$$

for all $-\infty < \xi < \infty$ and some $-\infty < \xi_0 < \infty$.

Proof. Setting $v = (u - \ell^+)/(\ell^- - \ell^+)$, the Burgers equation (9.6) can be transformed into the equation

$$v_t = v_{xx} + (b(v))_x \quad \text{where} \quad b(v) := -v\left\{\ell^+ + \frac{\ell^- - \ell^+}{2}v\right\}. \tag{9.7}$$

Furthermore, it can be verified that the Burgers equation admits a solution of the sought-after type if and only if (9.7) admits a wavefront solution from 1 to 0. However by Theorem 9.1, (9.7) has a solution of this type if and only if $b(v) > vb(1)$ for all $0 < v < 1$, in which case $\sigma = -b(1)$, and the solution g is given by

$$\ln\left\{\frac{1-g(\xi)}{g(\xi)}\right\} = \ln\left\{\frac{1-\nu}{\nu}\right\} + \frac{\ell^- - \ell^+}{2}\xi$$

for $-\infty < \xi < \infty$ for some $0 < \nu < 1$. The example may be completed by transferring these conclusions to the original equation. □

For a further illustration of Theorem 9.1, consider the foam drainage equation

$$u_t = (u^{3/2})_{xx} + (u^2)_x. \tag{9.8}$$

Application 9.3. *Equation* (9.8) *admits a global travelling-wave solution f with speed σ such that $f(\xi) \to \ell^{\pm}$ as $\xi \to \pm\infty$ for some nonnegative real numbers $\ell^- \neq \ell^+$ if and only if $\ell^- < \ell^+$ and $\sigma = -\ell^- - \ell^+$. In which case, for $\ell^- > 0$ the solution is given implicitly by*

$$\sqrt{\ell^+}\,\mathrm{arctanh}\sqrt{\frac{f(\xi)}{\ell^+}} - \sqrt{\ell^-}\,\mathrm{arctanh}\sqrt{\frac{\ell^-}{f(\xi)}} = \frac{\ell^+ - \ell^-}{3}(\xi - \xi_0), \tag{9.9}$$

and, for $\ell^- = 0$ explicitly by

$$f(\xi) = \begin{cases} 0 & \text{for } \xi \le \xi_0 \\ \ell^+ \tanh^2\left\{\frac{\sqrt{\ell^+}}{3}(\xi - \xi_0)\right\} & \text{for } \xi > \xi_0, \end{cases} \tag{9.10}$$

for all $-\infty < \xi < \infty$ and some $-\infty < \xi_0 < \infty$.

The proof of this example is similar to that of the previous one, and is omitted.

The Burgers equation (9.6) and the foam drainage equation (9.8) may both be viewed as special cases of the porous media equation with convection,

$$u_t = (u^m)_{xx} + b_0(u^n)_x. \tag{9.11}$$

Of note is that any nonnegative wavefront solution of the Burgers equation is necessarily positive everywhere, while the foam drainage equation admits a nonnegative wavefront solution whose support is bounded below. The next example indicates precisely when the porous media equation with convection admits a nonnegative wavefront solution whose support may be bounded on one side. Since the sign of b_0 is assumed to be arbitrary, without loss of generality attention is restricted to solutions with support bounded above.

Application 9.4. *Suppose that $m > 0$, $n > 0$ and b_0 and are real constants. Suppose furthermore that $0 < \ell < \infty$. Then equation* (9.11) *has a wavefront solution from ℓ to 0 if and only if $(n-1)b_0 < 0$, in which case the solution f necessarily has wave speed $\sigma = -b_0\ell^{n-1}$ and is given for all $-\infty < \xi < \infty$ and some $-\infty < \xi_0 < \infty$ by the following formulae.*

(a) *When $m > \min\{n, 1\}$:*

$$\frac{m}{b_0}\int_0^{f(\xi)} \frac{s^{m-2}}{s^{n-1} - \ell^{n-1}}\,ds = \max\{\xi_0 - \xi, 0\}. \tag{9.12}$$

(b) *When $m \le \min\{n, 1\}$:*

$$\frac{m}{b_0}\int_{f(\xi)}^{\ell/2} \frac{s^{m-2}}{s^{n-1} - \ell^{n-1}}\,ds = \xi - \xi_0. \tag{9.13}$$

Proof. The main conclusions may be read from Theorem 9.1. Furthermore, this theorem says that the distinct wavefront solution f of equation (9.11) is given modulo translation by $f(\xi) = 0$ for $\xi \geq \Xi_0$ and

$$\frac{m}{b_0} \int_{f(\xi)}^{\ell/2} \frac{s^{m-2}}{s^{n-1} - \ell^{n-1}} \, ds = \xi \quad \text{for } \xi < \Xi_0,$$

where

$$\Xi_0 := \frac{m}{b_0} \int_0^{\ell/2} \frac{s^{m-2}}{s^{n-1} - \ell^{n-1}} \, ds.$$

The formulae (9.12) and (9.13) may be obtained by computing that $\Xi_0 < \infty$ if and only if $m > \min\{n, 1\}$ and performing a suitable translation of the wavefront. □

The porous media equation with convection (9.11) incorporates the model of the flow of a thin viscous film over a flat plate previously discussed in Chapter 5 as a special case. In this model, u denotes the thickness of the fluid film, t time, x distance, b_0 the angle of inclination of the plate, $m = 4$ and $n = 3$. To fix ideas, let us imagine that x increases in a line running from left to right. Then, if the plate slopes upwards from left to right $b_0 > 0$, if the plate is horizontal $b_0 = 0$, and, if the plate slopes downwards $b_0 < 0$. Subsequently, in searching for a wavefront solution from ℓ to 0, we are looking for a profile in which the fluid film has a limiting thickness ℓ to the left and a limiting thickness 0 to the right. Application 9.4 shows that such a profile is possible if and only if the plate is inclined downhill from left to right. Moreover, the fluid film cannot cover the whole plate, and the flow must exhibit a leading edge which moves (like the rest of the profile) at a speed which is determined by the angle b_0 of the plate and the limiting thickness ℓ.

As a final illustration of the application of Theorem 9.1, we consider a model for the behaviour of a reactive solute in a porous medium. This model describes the transport of a single species dissolved in an incompressible fluid flowing through a homogeneous porous matrix with which the species reacts. If C denotes the concentration of the species per unit volume in the fluid, and S the concentration absorbed per unit mass on the matrix, then mass conservation implies

$$\frac{\partial}{\partial t}(\theta C + \rho S) + \operatorname{div} \mathbf{q} = 0$$

where t stands for time, θ the volumetric fluid content, ρ the bulk density of the matrix, and $\mathbf{q}$ the flux. The latter is viewed as being comprised of a diffusive component described by Fick's law and an advective component, which yields

$$\mathbf{q} = -D \operatorname{grad} C + C\mathbf{v}$$

where D denotes the coefficient of diffusivity and $\mathbf{v}$ the fluid flux. Additionally, assuming that the concentration of the species in the fluid and the concentration absorbed by the porous matrix are in equilibrium,

$$S = F(C)$$

for some fixed relation F known as the isotherm. The Langmuir and the Freundlich isotherms are the most well known. If finally the transport is supposed to be one-dimensional, and the fluid flow constant, combining the above equations and normalizing leads to

$$(u + F(u))_t = u_{xx} - ku_x \tag{9.14}$$

where u denotes the normalized concentration, $F(u)$ the rescaled isotherm and k the rescaled fluid flux.

Application 9.5. *Suppose that F is differentiable in $[\ell^-, \ell^+]$ with $F'(u) > -1$ for all $\ell^- < u < \ell^+$ for some $\ell^- < \ell^+$ and that k is a constant. Set*

$$G(u) := \frac{F(\ell^+)(u - \ell^-) + F(\ell^-)(\ell^+ - u)}{\ell^+ - \ell^-} - F(u)$$

and

$$m := 1 + \frac{F(\ell^+) - F(\ell^-)}{\ell^+ - \ell^-}.$$

Then the model (9.14) *admits a global travelling-wave solution f with speed σ such that $f(\xi) \to \ell^\pm$ as $\xi \to \pm\infty$ if and only if $kG(u) > 0$ for all $\ell^- < u < \ell^+$ and $\sigma = k/m$, in which case*

$$\int_{(\ell^+ + \ell^-)/2}^{f(\xi)} \frac{m}{kG(s)}\, ds = \xi - \xi_0 \quad \textit{for all } -\infty < \xi < \infty \tag{9.15}$$

and some $-\infty < \xi_0 < \infty$.

Proof. Define the function a on $[0, 1]$ by

$$\begin{aligned} F(\ell^+ - m\{\ell^+ - \ell^-\}a(v)) + \ell^+ - m\{\ell^+ - \ell^-\}a(v) \\ = \{F(\ell^+) + \ell^+\}(1 - v) + \{F(\ell^-) + \ell^-\}v, \end{aligned}$$

and note that $a(0) = 0$, $a'(v) > 0$ for all $0 < v < 1$, and $a(1) = 1/m$. Then by making the substitution $u = \ell^+ - m(\ell^+ - \ell^-)a(v)$, it can be verified that u is a solution of (9.14) if and only if v is a solution of the equation

$$v_t = (a(v))_{xx} - k(a(v))_x. \tag{9.16}$$

This last equation satisfies our basic hypotheses with coefficients defined on $[0, 1]$. Furthermore, a function f is a travelling-wave solution of (9.14) with the sought-after properties if and only if the function g defined via $f = \ell^+ - m(\ell^+ - \ell^-)a(g)$ is a wavefront solution of (9.16) from 1 to 0. By Theorem 9.1 though, equation (9.16) has such a solution if and only if

$$kv > kma(v) \quad \text{for all } 0 < v < 1, \tag{9.17}$$

in which case there is exactly one distinct such solution with speed $\sigma = k/m$, which modulo translation is given by

$$\int_{g(\xi)}^{\nu} \frac{ma'(s)}{ks - kma(s)}\, ds = \xi \quad \text{for } \Xi_1 < \xi < \Xi_0, \tag{9.18}$$

with $0 < \nu < 1$ and with Ξ_0 and Ξ_1 defined accordingly. Reformulating (9.17) in terms of the original equation justifies the necessary and sufficient condition for the existence of the travelling wave f; while reformulating (9.18) in the original variables gives the expression (9.15) for f in the interval $\Xi^- < \xi < \Xi^+$ where

$$\Xi^{\pm} = \xi_0 + \int_{(\ell^+ + \ell^-)/2}^{\ell^{\pm}} \frac{m}{kG(s)}\, ds$$

for some $-\infty < \xi_0 < \infty$. However, since F is differentiable in $[\ell^-, \ell^+]$ so too is G, and therefore $G(s) = G'(\ell^{\pm})(s - \ell^{\pm}) + o(s - \ell^{\pm})$ as $s \uparrow \ell^+$ and $s \downarrow \ell^-$ respectively. Whence it can be determined that $\Xi^{\pm} = \pm\infty$, and (9.15) holds for all ξ. □

Equation (9.14) also arises in analytical chemistry in the study of a platform atomizer when the interface processes are much more intense than the transport processes [233].

Bibliographic notes

A detailed discussion of the relation between solutions of (9.1) and (9.4) can be found in [129].

The class of explicit travelling-wave solutions of the Burgers equation presented in Application 9.2 was documented by the person after whom the equation is named in [62,63]. It may also be found in [134,156,157,187,234,262].

The explicit solution (9.10) of the foam drainage equation was recorded earlier in [131,292]; while the implicitly-defined solution (9.9) was documented previously in [293]. See also [304].

With regard to equation (9.14), we refer to [50] for a more detailed derivation of this equation as a model of a reactive solute in a porous medium. The ensuing analysis in Application 9.5 covers results of Rhee et al. [243], Bolt [50], and, van der Zee and Riemsdijk [313]. Extensions to situations in which the absorption processes is not an equilibrium process can be found in [86–91,132,244,245,271, 312], while an extension to the situation in which a simple reaction term is included can be found in [195]. Related results on the multi-dimensional equivalent of (9.14) in which the coefficients F and k have a periodic dependence on the spatial variables are included in [306–309]. The wavefront solution in Application 9.5 has been shown to be stable under certain conditions on F in [93,194].

Chapter 10

Wavefronts for reaction-diffusion

Throughout this chapter we consider only wavefront solutions of reaction-diffusion equations of the form

$$u_t = (a(u))_{xx} + c(u) \tag{10.1}$$

where a and c satisfy Hypothesis 2.1 with $\ell < \infty$. Since the pioneering work of Fisher [105] and of Kolmogorov, Petrovskii and Piskunov [186], much attention has been paid to the study of such solutions for equations of this class. The goal of this chapter is to show how the correspondence between travelling-wave solutions of equation (1.1) and solutions of the integral equation (1.9) may be invoked to generalize these earlier results. The theorems of Chapter 6 cover the previous results on semi-wavefront solutions of equations of the type (10.1).

For an equation such as the Nagumo equation, $u_t = u_{xx} + u(1-u)(u-\alpha)$ with $0 < \alpha < 1$, there are diverse possibilities for a wavefront solution, viz. from α to 0, from 0 to α, from 1 to α, from α to 1, from 0 to 1, and, from 1 to 0. To begin with, we shall consider these possibilities separately. In each case, by redefining the dependent variable one may regard the wavefront as one for the general equation $u_t = u_{xx} + c(u)$ from some ℓ to 0, where in the first four cases $c(u)$ has a fixed sign for $0 < u < \ell$ and in the last two cases $c(u)$ has one sign change for $0 < u < \ell$. As we shall see later, the results obtained under these conditions may be easily transferred to the various possibilities for a wavefront solution of the Nagumo equation. The same, of course, applies to other equations with a similar structure.

Our first result is one for which the primary conclusions are well-known for the semilinear version of equation (10.1).

Theorem 10.1. *Let $\ell < \infty$ and*

$$\kappa := \int_0^\ell c(s)a'(s)\,ds. \tag{10.2}$$

(a) *Suppose that $\kappa > 0$. Then equation (10.1) has a wavefront solution from ℓ to 0 with wave speed σ only if $\sigma > 0$ and*

$$\int_u^\ell c(s)a'(s)\,ds > 0 \quad \textit{for all } 0 < u < \ell.$$

(b) *Suppose that $\kappa = 0$. Then equation (10.1) has a wavefront solution from ℓ to 0 with wave speed σ if and only if $\sigma = 0$,*

$$\int_0^u c(s)a'(s)\,ds \le 0 \quad \textit{for all } 0 < u < \ell,$$

and

$$\theta(s) := \left| 2 \int_0^s c(r)a'(r)\,dr \right|^{1/2}$$

satisfies the integrability condition on $(0, \ell)$.

(c) *Suppose that* $\kappa < 0$. *Then equation* (10.1) *has a wavefront solution from* ℓ *to* 0 *with wave speed* σ *only if* $\sigma < 0$ *and*

$$\int_0^u c(s)a'(s)\,ds < 0 \quad \textit{for all } 0 < u < \ell.$$

The main conclusions are classically obtained directly from the ordinary differential equation $(a(f))'' + c(f) + \sigma f' = 0$ for a travelling-wave solution $u = f(x - \sigma t)$ of (10.1). Nonetheless, all the conclusions can also be easily deduced from the integral equation as follows. For a reaction-diffusion equation of the form (10.1), the integral equation (1.9) becomes

$$\theta(s) = \sigma s - \int_0^s \frac{c(r)a'(r)}{\theta(r)}\,dr. \tag{10.3}$$

Differentiating (10.3) gives

$$\theta'(s) = \sigma - \frac{c(s)a'(s)}{\theta(s)} \quad \text{for almost all } s. \tag{10.4}$$

Whence, multiplying by θ and integrating from 0 to s there holds

$$\frac{1}{2}\theta^2(s) = \sigma \int_0^s \theta(r)\,dr - \int_0^s c(r)a'(r)\,dr. \tag{10.5}$$

On the other hand, if θ is defined on $[0, \ell]$ and $\theta(\ell) = 0$, multiplying (10.4) by θ and integrating from s to ℓ there holds

$$-\frac{1}{2}\theta^2(s) = \sigma \int_s^\ell \theta(r)\,dr - \int_s^\ell c(r)a'(r)\,dr. \tag{10.6}$$

Therefore, since Theorem 2.10 states that (10.1) has a wavefront solution from ℓ to 0 with speed σ only if (10.3) has a solution θ satisfying the integrability condition on $[0, \ell]$ and $\theta(\ell) = 0$, we have that (10.5) and (10.6) necessarily hold for all $0 \le s \le \ell$ for such a function. In particular,

$$\sigma \int_0^\ell \theta(s)\,ds = \kappa. \tag{10.7}$$

The identities (10.5) – (10.7) give Theorem 10.1.

Theorem 10.1 can be demonstrated explicitly for a number of particular equations. Three will be examined below. Each contains a number of parameters that influence the sign of κ.

Application 10.2. *Consider the equation*

$$u_t = u_{xx} + \begin{cases} u(\ell^p - u^p)(u^p - \alpha^p) & \text{for } u > 0 \\ 0 & \text{for } u = 0 \end{cases} \tag{10.8}$$

with $0 < \alpha < \ell < \infty$.

(a) *When* $p > 0$ *equation* (10.8) *admits a wavefront solution from* ℓ *to* 0 *with speed* $\sigma = \{\ell^p - (p+1)\alpha^p\}/\sqrt{p+1}$ *and no other wave speed. The corresponding wavefront solution is given for some* $-\infty < \xi_0 < \infty$ *by*

$$f(\xi) = \ell \left[1 + \exp\left\{\frac{p}{\sqrt{p+1}}\ell^p(\xi - \xi_0)\right\}\right]^{-1/p} \tag{10.9}$$

for all $-\infty < \xi < \infty$.

(b) *When* $0 > p > -1$ *equation* (10.8) *admits a wavefront solution from* ℓ *to* 0 *with speed* $\sigma = \{(p+1)\alpha^p - \ell^p\}/\sqrt{p+1}$ *and no other wave speed. The corresponding wavefront solution is given for some* $-\infty < \xi_0 < \infty$ *by*

$$f(\xi) = \ell \left[1 - \exp\left(\frac{-p}{\sqrt{p+1}}\ell^p \min\{\xi - \xi_0, 0\}\right)\right]^{-1/p} \tag{10.10}$$

for all $-\infty < \xi < \infty$.

Proof. By Theorem 8.3 part (vii), equation (10.8) has a wavefront solution from ℓ to 0 for at most one wave speed, and, by Theorem 8.7 for any wave speed the equation has at most one distinct wavefront solution. In the light of our integral equation theory, to confirm the main assertions of the example, it consequently suffices to show that for the stated value of σ the appropriate integral equation (10.3) admits a solution θ satisfying the integrability condition on $[0, \ell]$ and $\theta(\ell) = 0$. An easy computation verifies that $\theta(s) = s\,|\ell^p - s^p|/\sqrt{p+1}$ fits the bill. Hereafter, it is enough to note that the solutions (10.9) and (10.10) may be constructed explicitly following the procedure outlined in Section 2.3. □

For equation (10.8), $\kappa = p\ell^{p+2}\{\ell^p - (p+1)\alpha^p\}/\{2(p+1)(p+2)\}$. Thus with regard to Theorem 10.1 and the critical wave speed σ one can easily see that for $p\{\alpha - (p+1)^{-1/p}\ell\} < 0$ there holds $\sigma > 0$ and $\kappa > 0$, for $\alpha = (p+1)^{\ 1/p}\ell$ there holds $\sigma = 0$ and $\kappa = 0$, while for $p\{\alpha - (p+1)^{-1/p}\ell\} > 0$ there holds $\sigma < 0$ and $\kappa < 0$.

The case $p = 0$ is necessarily excluded from Application 10.2. For this case, equation (10.8) can be supplanted by its singular limit

$$u_t = u_{xx} + \begin{cases} u \ln(\ell/u) \ln(u/\alpha) & \text{for } u > 0 \\ 0 & \text{for } u = 0. \end{cases} \tag{10.11}$$

Application 10.3. *Equation* (10.11) *with* $0 < \alpha < \ell < \infty$ *admits a wavefront solution from* ℓ *to* 0 *with speed* $\sigma = \ln(\ell/\alpha) - 1$ *and no other wave speed. The corresponding wavefront solution is given for some* $-\infty < \xi_0 < \infty$ *by* $f(\xi) = \ell \exp\{-\exp(\xi - \xi_0)\}$ *for all* $-\infty < \xi < \infty$.

Proof. The logic behind the proof of this example is the same as that employed for the previous one. The solution of (10.3) in this case is $\theta(s) = s\ln(\ell/s)$. □

The value of (10.2) for Application 10.3 is $\kappa = \ell^2\{\ln(\ell/\alpha) - 1\}/4$. So, for this example also, the conclusion of Theorem 10.1 is evident. The wave speed σ and κ always have the same sign according to whether $\alpha < \ell/e$, $\alpha = \ell/e$ or $\alpha > \ell/e$.

A revealing illustration of Theorem 10.1 is provided by a model studied in [23]. This model is a simplified description of an exothermic heterogeneous reaction on a catalytic wire, such as the oxidation of ammonium on the surface of a wire of platinum. The basic ingredient of the model is the linear heat equation with a nonlinear source or sink term. This term is the net difference of two components. The first component is the heat energy generated per unit length of the wire by the reaction, and the other is the energy lost per unit length of wire to the environment. Supposing that by approximation the energy generation component may be described by a step-function, while the energy loss is given by a linear radiation condition, the net source or sink term becomes a piecewise linear function of the temperature with one discontinuity. Subsequently, normalizing the temperature of the environment to 0, the temperature at which the energy generation is in equilibrium with the radiation to ℓ, and the coefficient of thermal diffusivity in the equation to unity, the model takes the form of equation (10.12) below in which the unknown u denotes the normalized temperature.

Application 10.4. *The equation*

$$u_t = u_{xx} + \begin{cases} -u & \text{for } 0 \le u < \alpha \\ \ell - u & \text{for } \alpha \le u \le \ell \end{cases} \tag{10.12}$$

for some $0 < \alpha < \ell < \infty$ *admits a wavefront solution from* ℓ *to* 0 *with speed*

$$\sigma = \frac{\ell - 2\alpha}{\sqrt{\alpha(\ell - \alpha)}} \tag{10.13}$$

and no other wave speed. Moreover for some $-\infty < \xi_0 < \infty$ *the wavefront solution has the form*

$$f(\xi) = \begin{cases} \ell - (\ell - \alpha)\exp\left\{\sqrt{\frac{\alpha}{\ell-\alpha}}(\xi - \xi_0)\right\} & \text{for } \xi < \xi_0 \\ \alpha \exp\left\{\sqrt{\frac{\ell-\alpha}{\alpha}}(\xi_0 - \xi)\right\} & \text{for } \xi \ge \xi_0. \end{cases} \tag{10.14}$$

Proof. By Lemma 2.16(ii) with $\alpha = \beta = \sigma$ the appropriate equation of the form (10.3) has a unique solution

$$\theta(s;\sigma) := \frac{\sqrt{\sigma^2 + 4} + \sigma}{2} s$$

on $[0, \alpha]$. On the other hand, by the same token, the appropriate equation corresponding to (8.1) has a unique solution

$$\Theta(s; \sigma) := \frac{\sqrt{\sigma^2 + 4} - \sigma}{2} s$$

on $[0, \ell - \alpha]$. Following Lemma 8.9, we can subsequently find a solution of (10.3) satisfying the integrability condition on $[0, \ell]$ and taking the value 0 in ℓ if and only if we can find a σ such that $\theta(\alpha; \sigma) = \Theta(\ell - \alpha; \sigma) > 0$. It is easy to verify that this is possible if and only if σ takes the value (10.13). The wavefront solution f can be constructed accordingly. □

For this example one can compute $\kappa = \ell(\ell - 2\alpha)/2$. When $\alpha < \ell/2$ there holds $\kappa > 0$ and the travelling-wave solution of equation (10.12) has a positive speed σ. In [23] such a wavefront is termed an *ignition wave*. When $\alpha > \ell/2$ there holds $\kappa < 0$ and the wavefront has negative speed and is called an *extinction wave*. The stationary wave which occurs in the marginal case when $\alpha = \ell/2$ and $\kappa = 0$ is referred to as an *indifferent equilibrium*. This nomenclature is motivated by the asymptotic behaviour of the profile $u(x, t) = f(x - \sigma t)$ as $t \to \infty$. In the case of an ignition wave, $u(x, t) \to \ell$ as $t \to \infty$ for all x, and in the case of an extinction wave, $u(x, t) \to 0$ as $t \to \infty$ for all x, irrespective of any translation ξ_0. On the other hand, in the case $\kappa = 0$, there is an infinite number of possible steady states dependent upon the value of ξ_0 in (10.14). The inference is that it is the value of κ, which is prescribed solely by the reaction term, that determines the asymptotic behaviour of an arbitrary process.

10.1 Fixed sign

To proceed we consider the case of equation (10.1) in which $c(u)$ is positive for $0 < u < \ell$. Recall that via Theorem 3.4 any results for this case apply *mutatis mutandi* to the case that $c(u)$ is negative for $0 < u < \ell$.

Our main result is the following. A prominent feature when compared to the mainstream literature, is that no more continuity on c is required than that stated in Hypothesis 2.1.

Theorem 10.5 (Existence). *Suppose that $\ell < \infty$ and $c(u) > 0$ for all $0 < u < \ell$. Set*

$$\lambda_1 := \limsup_{r \downarrow 0} \left\{ \frac{1}{r} \int_0^r \frac{c(u)a'(u)}{u} \, du \right\}. \tag{10.15}$$

(a) *If $\lambda_1 = \infty$, equation (10.1) has no wavefront solution from ℓ to 0.*

(b) *If $\lambda_1 < \infty$, there exists a $\sigma^* > 0$ such that the equation has exactly one distinct wavefront solution from ℓ to 0 for every wave speed $\sigma \geq \sigma^*$ and no such solution for any wave speed $\sigma < \sigma^*$.*

Proof. By Theorem 2.10 and Lemma 3.5, equation (10.1) has a wavefront solution of the sought-after type if and only if (10.3) and the equation

$$\Theta(s) = -\sigma s + \int_0^s \frac{c(\ell - r)a'(\ell - r)}{\Theta(r)}\, dr \tag{10.16}$$

admit solutions satisfying the integrability condition on the interval $[0, \ell]$. However, by Lemma 2.16(i) equation (10.16) has a unique solution on $[0, \ell]$ which is positive on $(0, \ell)$ for every σ. On the other hand, by Lemma 6.6 equation (10.3) has a solution satisfying the integrability condition on $[0, \ell]$ for large σ if and only if $\Lambda_1(\ell) < \infty$, where

$$\Lambda_1(s) := \sup_{0<r<s} \left\{ \frac{1}{r} \int_0^r \frac{c(u)a'(u)}{u}\, du \right\}. \tag{10.17}$$

Plainly though $\Lambda_1(\ell) < \infty$ if and only if $\lambda_1 < \infty$. The theorem subsequently follows as a resumé of Theorems 8.3(i), 8.7(a) and 10.1. Moreover, in case (b), the critical wave speed σ^* can be characterized as the minimal value σ for which (10.3) has a solution on $[0, \ell]$. □

Equipped with two spatial variables, the equations

$$u_t = u_{xx} + u\,|\ln u| \tag{10.18}$$

and

$$u_t = u_{xx} - u\,|\ln u| \tag{10.19}$$

are exceptional in the group-theoretical classification of second-order parabolic equations [12, 112]. Regarding the first, we easily deduce the following from Theorem 10.5.

Application 10.6. *Equation* (10.18) *admits no wavefront solutions from* 1 *to* 0.

On the other hand, Theorems 3.4 and 10.5 tell us that there is a negative number such that (10.19) has exactly one distinct wavefront solution from 1 to 0 for every wave speed less than or equal to this number and no such solution for any wave speed greater than this number. It transpires that we can determine this critical number and the asymptotic behaviour of any wavefront solution of the equation explicitly. In the course of the remainder of this section, we shall validate the following.

Application 10.7. *Equation* (10.19) *admits exactly one distinct wavefront solution f from* 1 *to* 0 *for every wave speed $\sigma \le -2$ and no such solution for any wave speed $\sigma > -2$. Furthermore,*

$$\{1 - f(\xi)\}^{-1} f'(\xi) \to -\frac{2}{\sqrt{\sigma^2 - 4} - \sigma} \qquad \text{as } \xi \to -\infty$$

and

$$f^{-1}(\xi)\,|\ln f(\xi)|^{-1/2} f'(\xi) \to -1 \qquad \text{as } \xi \to \infty.$$

10.1.1 The critical wave speed

In the light of Theorem 8.2 the critical wave speed in Theorem 10.5 can be characterized as $\sigma^* = \min\{\mathcal{G}_s(\psi) : \psi \in \mathcal{R}\}$, where $\mathcal{R}$ denotes the set of nonnegative continuous functions ψ defined on I such that (8.2) holds, and $\mathcal{G}_s$ denotes the functional defined on $\mathcal{R}$ by (8.3). More substantially, the magnitude of the critical wave speed can be estimated in terms of the parameters λ_1 and Λ_1 given by (10.15) and (10.17) respectively,

$$\lambda_0 := \liminf_{r \downarrow 0} \left\{ \frac{1}{r} \int_0^r \frac{c(u)a'(u)}{u}\, du \right\},$$

$$\Lambda_0(s) := \inf_{0<r<s} \left\{ \frac{1}{r} \int_0^r \frac{c(u)a'(u)}{u}\, du \right\}$$

and

$$\Lambda_2 := \int_0^\ell \frac{c(s)a'(s)}{s^2}\, ds.$$

Note that $0 \le \Lambda_0(s) \le \lambda_0 \le \lambda_1 \le \Lambda_1(s) < \Lambda_2$ for all $0 < s \le \ell$. Define

$$\Phi(z_0, z_1) := \begin{cases} (2z_1 - z_0)/\sqrt{2(z_1 - z_0)} & \text{for } 0 \le 3z_0 < 2z_1 \\ 2\sqrt{z_0} & \text{for } 0 \le 2z_1 \le 3z_0 \end{cases} \tag{10.20}$$

and

$$\Phi^* := \sup\{\Phi(\Lambda_0(s), \Lambda_1(s)) : 0 < s < \ell\}.$$

Theorem 10.8 (Critical speed estimates). *Suppose that the hypotheses of Theorem* 10.5(b) *hold. Then*

$$\Phi^* \le \sigma^* \le 2\sqrt{\Lambda_1(\ell)}, \tag{10.21}$$

$$\sigma^* > \Phi^* \quad \text{if } \Phi^* > \Phi(\lambda_0, \lambda_1) \tag{10.22}$$

and

$$\sigma^* < \min\{2\sqrt{\Lambda_1(\ell)}, \sqrt{2\Lambda_2}\} \quad \text{if } \Lambda_1(\ell) > \lambda_1. \tag{10.23}$$

The next lemma will be used to prove this theorem.

Lemma 10.9. *Let the hypotheses of Theorem* 10.5(b) *hold.*

(i) *If $\sigma > \sigma^*$ or $\sigma \ge 2\sqrt{\Lambda_1(\ell)}$, any solution θ of* (10.3) *on* $[0, \ell]$ *satisfying $\theta(\ell) = 0$ cannot be the maximal solution of this equation.*

(ii) *If $\sigma = \sigma^* > 2\sqrt{\lambda_1}$, any solution θ^* of* (10.3) *on* $[0, \ell]$ *satisfying $\theta^*(\ell) = 0$ must be the maximal solution of this equation.*

Proof. (i) Let $\theta(\cdot; \sigma)$ denote the maximal solution of (10.3) on $[0, \ell]$ for each $\sigma \ge \sigma^*$. By Lemma 2.18(ii) we can estimate $\theta(s; \sigma) \ge (\sigma - \sigma^*)s + \theta(s; \sigma^*)$ for all $0 \le s \le \ell$. Subsequently when $\sigma > \sigma^*$ the maximal solution of (10.3) cannot vanish in ℓ. When $\sigma \ge 2\sqrt{\Lambda_1(\ell)}$, this conclusion is provided by Lemma 6.6 part (iv).

(ii) By Lemma 2.16(i), equation (10.16) has a unique solution $\Theta(\cdot;\sigma)$ on $[0,\ell]$ which is positive on $(0,\ell)$ for every σ. Thus by the proof of Lemma 3.5, $\theta^*(s) = \Theta(\ell - s;\sigma^*)$ for all $0 \le s \le \ell$; while, by Theorem 2.10, necessarily $\Theta(\ell;\sigma) > 0$ for every $\sigma < \sigma^*$. Now, choose $0 < \delta < \ell$ so small that $\Lambda_1(\delta) < (\sigma^*)^2/4$. By Lemma 6.6(iv), for every $\sigma \ge 2\sqrt{\Lambda_1(\delta)}$ equation (10.3) has a maximal solution $\theta(\cdot;\sigma)$ on $[0,\delta]$ such that

$$\theta(s;\sigma) \ge \frac{\sigma + \sqrt{\sigma^2 - 4\Lambda_1(\delta)}}{2}s \quad \text{for all } 0 < s < \delta. \tag{10.24}$$

Subsequently, since for any $\sigma^* > \sigma > 2\sqrt{\Lambda_1(\delta)}$ both $\theta(\cdot;\sigma)$ and $\Theta(\ell - s;\sigma)$ satisfy the equation

$$\theta(s) = \theta(0) + \sigma s - \int_0^s \frac{c(r)a'(r)}{\theta(r)}\,dr$$

on $[0,\delta]$, while $\Theta(\ell;\sigma) > \theta(0;\sigma)$, applying Lemma 2.18(ii) there holds

$$\Theta(\ell - s;\sigma) > \theta(s;\sigma) \quad \text{for all } 0 \le s \le \delta. \tag{10.25}$$

Combining (10.24) and (10.25), and invoking Lemma 8.5 for equation (10.16) to justify passage to the limit $\sigma \uparrow \sigma^*$, yields (10.24) with θ^* and σ^* in place of $\theta(\cdot;\sigma)$ and σ respectively. By Lemmata 6.6(iv) and 6.8(i) though, equation (10.3) with $\sigma = \sigma^*$ admits at most one solution on $[0,\delta)$ satisfying such an inequality, namely its maximal solution $\theta(\cdot;\sigma^*)$. Hence, $\theta^* \equiv \theta(\cdot;\sigma^*)$ on $[0,\delta)$. Hereafter, by Lemma 4.7, θ^* has to be the maximal solution of (10.3) with $\sigma = \sigma^*$ on the whole of $[0,\ell]$. □

Proof of Theorem 10.8. Recalling that σ^* is essentially the minimum value σ such that (10.3) has a solution on $[0,\ell]$, Lemma 6.6 parts (ii) and (iv) give (10.21). To confirm (10.22) we note that Φ is nondecreasing in both its arguments. In particular, $\Phi(\lambda_0,\lambda_1) \ge \Phi(\Lambda_0(s),\lambda_1) = \Phi(\Lambda_0(s),\Lambda_1(s))$ for all $0 < s \le \ell$ such that $\Lambda_1(s) = \lambda_1$, and, $\Phi(\lambda_0,\lambda_1) \ge 2\sqrt{\lambda_0} \ge 2\sqrt{\Lambda_0(s)} = \Phi(\Lambda_0(s),\Lambda_1(s))$ for all $0 < s \le \ell$ such that $3\Lambda_0(s) \ge 2\Lambda_1(s)$. Consequently, if $\Phi^* > \Phi(\lambda_0,\lambda_1)$ there exists a $0 < s^* \le \ell$ such that $\Phi^* = \Phi(\Lambda_0(s^*),\Lambda_1(s^*))$, $\Lambda_1(s^*) > \lambda_1$, and, $3\Lambda_0(s^*) < 2\Lambda_1(s^*)$. Lemma 6.6 part (iii) then provides (10.22). With regard to (10.23), Lemma 10.9 parts (i) and (ii) preclude equality on the right-hand side of (10.21) when $\Lambda_1(\ell) > \lambda_1$. It therefore suffices to show that $\sigma < \sqrt{2\Lambda_2}$ whenever $\Lambda_2 < \infty$. To achieve this, without risk of ambiguity, we drop the asterisk from the notation of σ^* and thereafter let θ denote the unique solution of equation (10.3) on $[0,\ell]$ such that $\theta(\ell) = 0$. Noting that if $\Lambda_2 < \infty$, necessarily $\lambda_0 = \lambda_1 = 0$, Lemmata 6.6, 6.8 and 10.9(ii) imply that $\theta(s)/s \to \sigma$ as $s \downarrow 0$. Simultaneously, multiplying (10.4) by $\theta(s)/s$ there holds

$$\left(\frac{\theta^2}{s^2}\right)' = 2\frac{\{\sigma s - \theta(s)\}\theta(s)}{s^3} - 2\frac{c(s)a'(s)}{s^2} \quad \text{for almost all } 0 < s < \ell.$$

So, integrating from 0 to ℓ, we have

$$\sigma^2 = 2\int_0^\ell \frac{c(s)a'(s)}{s^2}\,ds - 2\int_0^\ell \frac{\{\sigma s - \theta(s)\}\theta(s)}{s^3}\,ds.$$

This provides the outstanding inequality. □

Previous estimates of the critical speed σ^* in Theorem 10.5(b) have been mainly concerned with the semilinear version of (10.1) with a differentiable reaction term. Applying these results to the general equation (10.1) under the assumption that ca' is differentiable, they read

$$2\sqrt{(ca')'(0)} \le \sigma^* \le 2\sqrt{\sup_{0<u<\ell} \frac{c(u)a'(u)}{u}} \le 2\sqrt{\sup_{0<u<\ell} (ca')'(u)}. \tag{10.26}$$

We contend that Theorem 10.8 covers and improves these estimates. This can be discerned from the following.

Corollary 10.10 (of Theorem 10.8). *Suppose that $\ell < \infty$, $c(u) > 0$ for all $0 < u < \ell$, ca' is differentiable in 0, and, $(ca')(0) = 0$. Then there exists a $\sigma^* > 0$ such that equation* (10.1) *admits precisely one distinct wavefront solution from ℓ to 0 for every wave speed $\sigma \ge \sigma^*$ and no such solution for any wave speed $\sigma < \sigma^*$. Moreover:*

(a) *If $\Lambda_1(\ell) = (ca')'(0)$, then $\sigma^* = 2\sqrt{(ca')'(0)}$.*

(b) *If $\Lambda_1(\ell) > (ca')'(0)$, then $\sigma^* \ge 2\sqrt{(ca')'(0)}$, $\sqrt{2\Lambda_1(\ell)} < \sigma^* < 2\sqrt{\Lambda_1(\ell)}$, and,*

$$\Lambda_1(\ell) < \sup_{0<u<\ell} \frac{c(u)a'(u)}{u}. \tag{10.27}$$

Proof. In the present situation it can be verified that $\lambda_0 = \lambda_1 = (ca')'(0)$ and hence $\Phi(\lambda_0, \lambda_1) = 2\sqrt{(ca')'(0)}$. Moreover, $\Phi(\Lambda_0(\ell), \Lambda_1(\ell)) > \Phi(0, \Lambda_1(\ell)) = \sqrt{2\Lambda_1(\ell)}$ if $\Lambda_0(\ell) > 0$, while $\Phi(\Lambda_0(\ell), \Lambda_1(\ell)) = \sqrt{2\Lambda_1(\ell)} > 0 = \Phi(\lambda_0, \lambda_1)$ if $\Lambda_0(\ell) = 0$. Thus the estimates of the magnitude of σ^* are easily derived from Theorem 10.8. The only element of this corollary which requires further justification is (10.27) in case (b). This inequality without the strictness is deducible by elementary analysis. The strictness follows from the observation that equality would imply that $c(u)a'(u) \le (ca')'(0)u$ for all $0 \le u \le \ell$; in which event, case (a) would apply. □

In general, the inequality $\sigma^* \ge 2\sqrt{(ca')'(0)}$ in part (b) will not be strict [7, 150].

The well-known results for the KPP equation on the existence of wavefront solutions are a simple consequence of Corollary 10.10. The critical wave speed asserted in Application 10.7 is also given by this corollary when one first takes Theorem 3.4 into consideration.

In the study of the stability properties of travelling-wave solutions of the semilinear version of equation (10.1),

$$u_t = u_{xx} + c(u),$$

Stokes [276] made a distinction between wavefront solutions with the critical wave speed σ^* according to whether or not $\sigma^* = 2\sqrt{c'(0)}$. If $\sigma^* = 2\sqrt{c'(0)}$ Stokes calls such a wavefront solution a *pulled* wave, whereas if $\sigma^* > 2\sqrt{c'(0)}$ this wave is called a *pushed* wave. These designations are motivated by the observation that in the first case the critical speed is determined as it were by the behaviour of $c(u)$ as $u \downarrow 0$ and this is reflected in the front (or pulling edge) of the wave $f(\xi)$ as $\xi \to \infty$. On the other hand, in the second case the critical speed is also influenced by the behaviour of $c(u)$ for $u > 0$ and this is reflected in the body of the wave f (pushing from behind). Lemma 10.9 implies that in the sense of Stokes [276] any pushed wave necessarily corresponds with a maximal solution of the integral equation (10.3). A pulled wave may or may not correspond with such a solution. In the concluding paragraph of [276], Stokes remarks that if the wavefront solution with the critical speed $\sigma^* = 2\sqrt{c'(0)}$ corresponds with a maximal solution of (10.3), then with respect to the stability properties this wave may be regarded as being both pushed and pulled. This corresponds with a hypothesis known in physics as the *marginal stability principle*. Other terms which have been used in this context are *linear front speed* and *Kolmogorov speed* for the wave speed $\sigma = 2\sqrt{c'(0)}$, *nonlinear front speed* for the critical wave speed $\sigma^* > 2\sqrt{c'(0)}$, *strongly heteroclinic front* for a wavefront solution associated with a maximal solution of the integral equation, *linear front* for a pulled wave, and, *nonlinear front* for a pushed wave [34, 38, 77, 133, 241, 242].

As a simple illustration of Theorems 10.5 and 10.8, let us consider the following studied earlier in [153, 305], where the specific interest was in the unusual behaviour of the wavefront profile in the limit $\Delta \downarrow 0$.

Application 10.11. *The equation*

$$u_t = (D(u)u_x)_x + u(1-u)$$

where

$$D(u) = \frac{1+\omega}{2} + \frac{1-\omega}{2}\tanh\left(\frac{u-u_0}{\Delta}\right)$$

for some parameters $0 < \omega < 1$, $0 < u_0 < 1$ and $\Delta > 0$ admits a wavefront solution from 1 to 0 with speed σ if and only if $\sigma \geq \sigma^*$ for some value σ^* with

$$2\sqrt{D(0)} \leq \sigma^* < 2\sqrt{D(1)}. \tag{10.28}$$

Since $D(0) > w$ and $D(1) < 1$, the estimate (10.28) sharpens the equivalent one in [153].

Proof. In this example $c(u)a'(u) = u(1-u)D(u)$. Thus $(ca')'(0) = D(0)$ and since D is strictly increasing on $[0, \ell]$ we can estimate $\Lambda_1(1) < D(1)$. Theorem 10.5 and Corollary 10.10 give the result from these estimates. □

Unless of course $2\sqrt{\Lambda_1(\ell)} = \Phi^*$, Theorem 10.8 does not give the critical wave speed σ^* exactly. However if luck would have it that for a certain value of σ, equation (10.3) can be shown to admit an explicit solution and this explicit solution possesses certain properties, then σ^* can be identified. This assertion builds on Lemma 10.9 and is concretized below.

Theorem 10.12. *Suppose that the hypotheses of Theorem* 10.5(b) *hold. Suppose furthermore that for some number* $\widehat{\sigma} > 0$ *equation* (10.3) *with* $\sigma = \widehat{\sigma}$ *admits a solution* θ *on* $[0, \ell]$. *Then* $\sigma^* \leq \widehat{\sigma}$. *Moreover,* $\sigma^* = \widehat{\sigma}$ *in each of the following cases.*

(a) $\widehat{\sigma} = \Phi(\lambda_0, \lambda_1)$.

(b) $\widehat{\sigma} > 2\sqrt{\lambda_1}$, $\theta(\ell) = 0$, *and,*

$$\liminf_{s \downarrow 0} \frac{\theta(s)}{s} > \frac{\widehat{\sigma} - \sqrt{\widehat{\sigma}^2 - 4\lambda_1}}{2}. \tag{10.29}$$

Proof. Since following the proof of Theorem 10.5, σ^* is the minimal value of σ such that the integral equation (10.3) has a solution on $[0, \ell]$, necessarily $\sigma^* \leq \widehat{\sigma}$. Case (a) subsequently follows from the left-hand inequality in (10.21), as $\Phi^* \geq \Phi(\lambda_0, \lambda_1)$. To verify case (b), suppose, contrary to the conclusion, that $\sigma^* < \widehat{\sigma}$. Then, because $\theta(\ell) = 0$, by Lemma 10.9(i), θ cannot be the maximal solution of equation (10.3) with $\sigma = \widehat{\sigma}$. On the other hand, in view of (10.29), by Lemma 6.8 part (i), θ must be the maximal solution of (10.3) with $\sigma = \widehat{\sigma}$. So, by *reductio ad absurdum*, $\sigma^* < \widehat{\sigma}$ is excluded. □

The next two examples provide an illustration of the above theorem along with Theorems 10.5 and 10.8.

Consider a ferroelectric smetic-C^* liquid crystal subject to an electric field and a magnetic field parallel to the smetic layers and perpendicular to each other. Let u denote the azimuthal angle between the electric field and the polarization of the molecules in the crystal, $0 \leq u \leq \pi$. Then after a suitable rescaling, the dynamics of the twist angle are given by the equation

$$u_t = u_{xx} + \left\{ E - \left(E^2 + \frac{H^2}{4} \right) \cos u \right\} \sin u \tag{10.30}$$

where $0 < E \leq 1$ denotes the dimensionless electric field strength and H the dimensionless magnetic field strength [253, 275].

Application 10.13. *Suppose that* $0 < E \leq 1$ *and* $H^2 \leq 4E(1-E)$. *Then equation* (10.30) *admits exactly one distinct wavefront solution from* π *to* 0 *for every* *wave speed* $\sigma \geq \sigma^*$ *and no such solution for any wave speed* $\sigma < \sigma^*$, *where*

$$\sigma^* = \begin{cases} \sqrt{4E(1-E) - H^2} & \text{if } H^2 < 2E(1-2E) \\ 2E/\sqrt{4E^2 + H^2} & \text{if } H^2 \geq 2E(1-2E). \end{cases}$$

Proof. Let

$$\theta_0 := \sqrt{E^2 + \frac{H^2}{4}}$$

and note that $\theta_0^2 \leq E$. Then computing that $\lambda_0 = \lambda_1 = E - \theta_0^2$, there holds $\sigma^* \geq \Phi(\lambda_0, \lambda_1) = 2\sqrt{E - \theta_0^2}$ by Theorem 10.8. The key to the full result is that the integral equation (10.3) admits the explicit solution

$$\widehat{\theta}(s) := \theta_0 \sin s$$

for the wave speed

$$\widehat{\sigma} := \frac{E}{\theta_0}.$$

Thus, by Theorem 10.12, we also have $\sigma^* \leq \widehat{\sigma}$. To proceed we distinguish three cases. These correspond to $H^2 = 2E(1-2E)$, $H^2 > 2E(1-2E)$ and $H^2 < 2E(1-2E)$ respectively.

(a) $\theta_0^2 = E/2$. In this case, the estimates $\sigma^* \geq 2\sqrt{E - \theta_0^2}$ and $\sigma^* \leq \widehat{\sigma}$ coincide.

(b) $\theta_0^2 > E/2$. There holds

$$\lim_{s \downarrow 0} \frac{\widehat{\theta}(s)}{s} = \theta_0 > \frac{E - \theta_0^2}{\theta_0} = \frac{\widehat{\sigma} - \sqrt{\widehat{\sigma}^2 - 4\lambda_1}}{2}.$$

So, by Theorem 10.12(b), $\sigma^* = \widehat{\sigma}$.

(c) $\theta_0^2 < E/2$. Defining $\widetilde{\theta}(s) := \theta(s)/\sqrt{E - \theta_0^2}$ for $0 \leq s \leq \pi$ and $\widetilde{\sigma} := \sigma/\sqrt{E - \theta_0^2}$, the integral equation can be rescaled so that $\widetilde{\theta}$ becomes the unknown, σ is replaced by $\widetilde{\sigma}$, and, the reaction term becomes

$$\widetilde{c}(u) := \frac{E - \theta_0^2 \cos u}{E - \theta_0^2} \sin u.$$

Let $\widetilde{\sigma}^*$ denote the least value of $\widetilde{\sigma}$ for which this rescaled integral equation has a solution on $[0, \pi]$. Now, $\widetilde{c}(u)$ is nondecreasing as a function of θ_0^2 for $0 \leq u \leq \pi$. From our very first estimate we know that $\widetilde{\sigma}^* \geq 2$. On the other hand, from part (a) we know that $\widetilde{\sigma}^* = 2$ when $\theta_0^2 = E/2$. Hence, by Theorem 8.4(b), we must have $\widetilde{\sigma}^* = 2$ for $0 < \theta_0^2 < E/2$. Returning to the original formulation of the integral equation gives the result. ☐

For liquid crystal displays, the critical wave speed σ^* is an important parameter with regard to the switching time.

The next application includes the Fisher equation, the Newell–Whitehead equation and the Zeldovich equation, all of which may be considered as benchmarks for the study of the general equation (10.1).

Application 10.14. *The equation*

$$u_t = u_{xx} + u(\ell^p - u^p)(c_0\ell^p + c_1 u^p), \tag{10.31}$$

where $\ell > 0$, $p > 0$, and, c_0 and c_1 are real constants such that $c_0\ell^p + c_1u^p > 0$ for all $0 < u < \ell$, admits exactly one distinct wavefront solution from ℓ to 0 for every wave speed $\sigma \geq \sigma^*$ and no such solution for any wave speed $\sigma < \sigma^*$, where

$$\sigma^* = \begin{cases} 2\sqrt{c_0}\ell^p & \text{if } c_1 \leq (p+1)c_0 \\ \dfrac{(p+1)c_0 + c_1}{\sqrt{(p+1)c_1}}\ell^p & \text{if } c_1 > (p+1)c_0. \end{cases}$$

Proof. The justification of this example is similar to that of the previous one. We compute that $\lambda_0 = \lambda_1 = c_0\ell^{2p}$. Hence, using Theorem 10.8, $\sigma^* \geq \Phi(\lambda_0, \lambda_1) = 2\sqrt{c_0}\ell^p$. The full result is obtained from the observation that when $c_1 > 0$ and

$$\widehat{\sigma} := \frac{(p+1)c_0 + c_1}{\sqrt{(p+1)c_1}}\ell^p,$$

the integral equation (10.3) with $\sigma = \widehat{\sigma}$ admits the explicit solution

$$\widehat{\theta}(s) := \sqrt{\frac{c_1}{p+1}}s(\ell^p - s^p).$$

So, by Theorem 10.12, $\sigma^* \leq \widehat{\sigma}$ for $c_1 > 0$. There are now three cases to be distinguished.

(a) $c_1 = (p+1)c_0$. In this case, the assertion is immediate from the estimates already obtained.

(b) $c_1 > (p+1)c_0$. In this case there holds

$$\lim_{s\downarrow 0}\frac{\widehat{\theta}(s)}{s} = \sqrt{\frac{c_1}{p+1}}\ell^p > \sqrt{\frac{p+1}{c_1}}c_0\ell^p = \frac{\widehat{\sigma} - \sqrt{\widehat{\sigma}^2 - 4\lambda_1}}{2}.$$

So, by Theorem 10.12(b), $\sigma^* = \widehat{\sigma}$.

(c) $c_1 < (p+1)c_0$. By Theorem 8.4(b), σ^* is a nondecreasing function of c_1. Using case (a) we therefore deduce that $\sigma^* \leq 2\sqrt{c_0}\ell^p$. In combination with the earlier deduction that the reverse inequality must hold, this determines σ^* in this final case. □

The consequences of Application 10.14 for the benchmark equations mentioned earlier are well-known.

- The Fisher equation, $u_t = u_{xx} + u(1-u)$, admits precisely one distinct wavefront solution from 1 to 0 for every wave speed $\sigma \geq 2$ and no such solution for any wave speed $\sigma < 2$.

- The Newell–Whitehead equation, $u_t = u_{xx} + u(1-u^2)$, likewise admits precisely one distinct wavefront solution from 1 to 0 for every wave speed $\sigma \geq 2$ and no such solution for any wave speed $\sigma < 2$.
- The Zeldovich equation, $u_t = u_{xx} + u^2(1-u)$, admits precisely one distinct wavefront solution from 1 to 0 for every wave speed $\sigma \geq 1/\sqrt{2}$ and no such solution for any wave speed $\sigma < 1/\sqrt{2}$.

For all three of the above equations one needs to take the parameter values $\ell = p = 1$ in Application 10.14. In addition, for the Fisher equation $c_0 = 1$ and $c_1 = 0$, for the Newell–Whitehead $c_0 = c_1 = 1$, and, for the Zeldovich equation $c_0 = 0$ and $c_1 = 1$.

Without going into the nitty-gritty of the calculations, we note the following consequences of the explicit determination of the critical speed σ^* in Application 10.14:

$$\sigma^* = \Phi(\lambda_0, \lambda_1) \quad \text{when } c_1 \leq (p+1)c_0,$$

$$\sigma^* = 2\sqrt{\Lambda_1(\ell)} \quad \text{when } c_1 \leq c_0,$$

and,

$$\sigma^* \sim \Phi^* \sim \sqrt{2\Lambda_2} \quad \text{as } p \to \infty \quad \text{when } c_0 = 0.$$

Thus, in general, the estimates in (10.21) and (10.23) are sharp. Furthermore, since $\Lambda_1(\ell) = \lambda_1$ if and only if $c_1 \leq c_0$, the condition $\Lambda_1(\ell) > \lambda_1$ may be viewed as both necessary and sufficient for strict inequality on the right-hand side of (10.21). Whereas this condition can be seen not to be sufficient for strictness on the left-hand side of (10.21). This is further confirmation that $\sigma^* = 2\sqrt{(ca')'(0)}$ cannot be excluded from the conclusions of part (b) of Corollary 10.10. Finally, because $\Phi(\lambda_0, \lambda_1) = \Phi^* < \sigma^*$ when $(p+1)c_0 < c_1 \leq (p+3)c_0$ — and the last inequality is not sharp — it can be deduced that although the condition $\Phi^* > \Phi(\lambda_0, \lambda_1)$ is sufficient for strictness on the left-hand side of (10.21), it is not necessary.

In the past, the critical wave speed has been identified in a number of specific instances of Application 10.14 using the marginal stability principle. Combination of the marginal stability principle together with results in [240] lead Goriely [133] to a conjecture for the magnitude of the critical wave speed when the reaction term on the right-hand side of (10.31) is replaced by any polynomial which is positive for $0 < u < \ell$ and vanishes in 0 and ℓ. The focus was on the equation

$$u_t = u_{xx} + k\ell^{m-1}u + (1-k)\ell^{m-n}u^n - u^m \tag{10.32}$$

where $m > n > 1$ are integers and $0 < k < 1$. For this particular equation, the conjectured critical wave speed is

$$\sigma_G := \begin{cases} \dfrac{2(m-n) + (m+1)(n-1)k}{2\sqrt{\gamma(1+\gamma)\{m-n+(n-1-\gamma)k\}}}\ell^{(m-1)/2} & \text{if } k < k_G \\[2ex] 2\sqrt{k}\ell^{(m-1)/2} & \text{if } k \geq k_G, \end{cases}$$

where

$$\gamma := \sqrt{\frac{(m-1)(n-1)}{2}} \quad \text{and} \quad k_G := \frac{2(m-n)}{(m-3)(n-1)+4\gamma}.$$

Application 10.14 with $p = (m-1)/2$, $c_0 = k$ and $c_1 = 1$ shows that σ_G is indeed the critical wave speed when $n = (m+1)/2$. For a reaction term comprising a specific polynomial with five terms, the conjecture was refuted in general in [70]. The following application of Corollary 10.10 shows that it is also not true in general when restricted to equations of the type (10.32).

Application 10.15. *Let $k \geq 0$, $\ell > 0$ and $m > n > 1$ be real parameters. Then there exists a $\sigma^* > 0$ such that equation (10.32) admits precisely one distinct wavefront solution from ℓ to 0 for every wave speed $\sigma \geq \sigma^*$ and no such solution for any wave speed $\sigma < \sigma^*$. Moreover:*

(a) *If $k < 1$ there holds $4K > (\sigma^*)^2\ell^{1-m} > \Phi^2(k, K)$, where Φ is given by (10.20) and*

$$K := k + \frac{m-n}{m(n-1)}\left\{\frac{m(n-1)}{n(m-1)}(1-k)\right\}^{(m-1)/(m-n)}.$$

(b) *If $k \geq 1$ there holds $(\sigma^*)^2\ell^{1-m} = 4k$.*

Proof. Rewriting equation (10.32) as $u_t = u_{xx} + k\ell^{m-n}u(\ell^{n-1} - u^{n-1}) + u^n(\ell^{m-1} - u^{m-1})$, it can be seen to satisfy the prerequisites of Theorem 10.5. Furthermore, one can compute that $\lambda_1 = \lambda_0 = k\ell^{m-1}$, that $\Lambda_1(\ell) = \lambda_1$ when $k \geq 1$, and, that there exists a $0 < s^* < \ell$ such that $\Lambda_1(\ell) = \Lambda_1(s^*) = K\ell^{m-1} > \Lambda_0(s^*) = \lambda_0$ when $k < 1$. Application of Theorems 10.5 and 10.8 subsequently yields the result. □

To refute the conjecture of Goriely, we merely note that when $n = 2$ and $0 < k < 1$ there holds $(\sigma_G)^2\ell^{1-m} - 4K \to k^2/2$ as $m \to \infty$. Thus for large m, even under the constraint that m is an integer, the conjectured critical speed σ_G violates a proven upper bound for the true critical wave speed σ^*.

In general, additional useful information on the critical speed can be obtained from the following three theorems.

Theorem 10.16. *Consider equation (10.1) with two different sets of coefficients a_1, c_1 and a_2, c_2 such that the conditions of Theorem 10.5(b) hold on the same interval $[0, \ell]$. Let κ_i, $\lambda_1^{(i)}$, and σ_i^* denote the appropriate parameters associated with (10.1) for $i = 1, 2$. Then if $c_1(u)a_1'(u) \geq c_2(u)a_2'(u)$ for all $0 < u < \ell$, there holds $\sigma_1^* \geq \sigma_2^*$. Moreover, if $\kappa_1 > \kappa_2$ and*

$$\sigma_1^* > 2\sqrt{\lambda_1^{(2)}}, \tag{10.33}$$

there holds $\sigma_1^* > \sigma_2^*$.

Proof. The first assertion is a restatement of Theorem 8.4(b). To prove the second assertion, suppose to its contrary that $\sigma_1^* = \sigma_2^*$. Let θ_i denote the solution of (10.3) with $\sigma = \sigma_i^*$ and $c(r)a'(r) = c_i(r)a_i'(r)$ such that θ_i satisfies the integrability condition on $[0,\ell]$ and $\theta_i(\ell) = 0$ for $i = 1, 2$. By Lemma 2.18, equation (10.3) with $\sigma = \sigma_2^*$ and $c(r)a'(r) = c_2(r)a_2'(r)$ has a maximal solution θ_2^* on $[0,\ell]$ such that $\theta_2^*(s) \geq \max\{\theta_1(s), \theta_2(s)\}$ for all $0 \leq s \leq \ell$. Moreover, by Lemma 4.7, either $\theta_2^* \equiv \theta_2$ or $\theta_2^*(\ell) > \theta_2(\ell)$. The latter alternative however is ruled out by Lemma 10.9(ii). So, $\theta_2(s) = \theta_2^*(s) \geq \theta_1(s)$ for all $0 \leq s \leq \ell$. Recalling (10.7) this implies $\kappa_2 \geq \kappa_1$. Hence, by *reductio ad absurdum*, the assertion must be true. □

Note that on its own the assumption $\kappa_1 > \kappa_2$ is insufficient to guarantee strict inequality in the conclusion of this theorem. In fact, even the assumption that $c_1(u)a_1'(u) > c_2(u)a_2'(u)$ for all $0 < u < \ell$ is inadequate. This is demonstrated by Application 10.14. For fixed $0 < \ell < \infty$, $p > 0$ and $c_0 > 0$, the source term in equation (10.31) is a strictly increasing function of $c_1 \geq -c_0$ for every $0 < u < \ell$. However, σ^* increases strictly with c_1 if and only if $c_1 \geq (p+1)c_0$. Moreover, noting that for equation (10.31) one has $\sigma^* = 2\sqrt{\lambda_1}$ if and only if $c_1 \leq (p+1)c_0$, this example shows that, in general, the condition (10.33) is both necessary and sufficient for strict inequality in the conclusion of the theorem.

Theorem 10.17. *Consider equation* (10.1) *with a fixed set of coefficients and with a sequence of coefficients* $\{a_n, c_n\}_{n=1}^{\infty}$ *such that the conditions of Theorem* 10.5(b) *hold on the same interval* $[0,\ell]$. *Let* σ^* *and* $\{\sigma_n^*\}_{n=1}^{\infty}$ *denote the corresponding critical wave speeds. Define*

$$\widetilde{c}_n(u)\widetilde{a}_n'(u) := \sup\{c_j(u)a_j'(u) : j \geq n\} \quad \textit{for } 0 < u < \ell,$$

and, let $\widetilde{\lambda}_1^{(n)}$ *denote the parameter defined by the relations* (10.17) *and* (10.15) *with* $c(u)a'(u)$ *replaced by* $\widetilde{c}_n(u)\widetilde{a}_n'(u)$. *Suppose that* $c_n a_n' \to ca'$ *in* $L^1(0,\ell)$ *as* $n \to \infty$. *Then* $\liminf_{n\to\infty} \sigma_n^* \geq \sigma^*$. *Moreover, if* $\widetilde{c}_n\widetilde{a}_n' \in L^1(0,\ell)$ *for large* n *and*

$$\sigma^* \geq \lim_{n\to\infty} 2\sqrt{\widetilde{\lambda}_1^{(n)}}, \tag{10.34}$$

there holds $\sigma_n^* \to \sigma^*$ *as* $n \to \infty$.

Proof. For the proof of the main assertion, we do not need the functions $\widetilde{c}_n\widetilde{a}_n'$ from the statement of the theorem. We therefore introduce no ambiguity if we temporarily redefine $\widetilde{c}_n(s)\widetilde{a}_n'(s) := \inf\{c_j(s)a_j'(s) : j \geq n\}$ for $0 < s < \ell$. Then, by hypothesis, $\widetilde{c}_n\widetilde{a}_n' \to ca'$ in $L^1(0,\ell)$ as $n \to \infty$. Furthermore, $\widetilde{c}_n(s)\widetilde{a}_n'(s) \leq \widetilde{c}_{n+1}(s)\widetilde{a}_{n+1}'(s) \leq c(s)a'(s)$ and $\widetilde{c}_n(s)\widetilde{a}_n'(s) \leq c_n(s)a_n'(s)$ for almost all $0 < s < \ell$ and all $n \geq 1$. Moreover, although equation (10.1) with a and c replaced by $\widetilde{a}_n$ and $\widetilde{c}_n$ might not satisfy Hypothesis 2.1 nor the hypotheses of Theorem 10.5(b), the equation

$$\theta_n(s) = \sigma s - \int_0^s \frac{\widetilde{c}_n(r)\widetilde{a}_n'(r)}{\theta_n(r)}\,dr \tag{10.35}$$

does fit into the integral equation theory presented in Section 2.2. Thus, recalling that σ^* and σ_n^* are essentially the minimum values of σ such that (10.3) and

$$\theta(s) = \sigma s - \int_0^s \frac{c_n(r)a_n'(r)}{\theta(r)}\, dr \tag{10.36}$$

respectively have a solution on $[0, \ell]$, if we define $\widetilde{\sigma}_n^*$ as the minimum value of σ such that (10.35) has a solution on $[0, \ell]$, by Lemma 2.18 there holds $\widetilde{\sigma}_n^* \le \widetilde{\sigma}_{n+1}^* \le \sigma^*$ and $\widetilde{\sigma}_n^* \le \sigma_n^*$ for all $n \ge 1$. Set $\widetilde{\sigma}_\infty^* := \lim_{n\to\infty} \widetilde{\sigma}_n^*$. Now let $\sigma > \widetilde{\sigma}_\infty^*$ be arbitrary. Then for each $n \ge 1$, equation (10.35) has a maximal solution θ_n on $[0, \ell]$, and, by Lemma 2.18 there holds $\theta_n(s) \ge \theta_{n+1}(s)$ for all $0 \le s \le \ell$ and $n \ge 1$. Whence, $\widetilde{c}_n(r)\widetilde{a}_n'(r)/\theta_n(r) \le \widetilde{c}_{n+1}(r)\widetilde{a}_{n+1}'(r)/\theta_{n+1}(r)$ for almost all $0 < r < \ell$ and $n \ge 1$. So by the Monotone Convergence Theorem, if we define $\theta_\infty(s) := \lim_{n\to\infty} \theta_n(s)$ for all $0 \le s \le \ell$, we may let $n \to \infty$ in (10.35) to deduce that θ_∞ satisfies (10.3) on $[0, \ell]$. This infers that $\sigma \ge \sigma^*$. Hence, in view of the arbitrariness of σ, we have $\widetilde{\sigma}_\infty^* \ge \sigma^*$. Since, $\widetilde{\sigma}_\infty^* \le \liminf_{n\to\infty} \sigma_n^*$, this proves the main assertion of the theorem. To confirm the remaining assertion, we restore $\widetilde{c}_n\widetilde{a}_n'$ to its original definition. Without loss of generality we assume that $\widetilde{c}_n\widetilde{a}_n' \in L^1(0, \ell)$ and $\widetilde{\lambda}_1^{(n)} < \infty$ for all $n \ge 1$. Subsequently, arguing as above, we can once more define $\widetilde{\sigma}_n^*$ as the minimum value of σ such that (10.35) has a solution on $[0, \ell]$, and, conclude that $\widetilde{\sigma}_n^* \ge \widetilde{\sigma}_{n+1}^* \ge \sigma^*$ and $\widetilde{\sigma}_n^* \ge \sigma_n^*$ for all $n \ge 1$. The proof of the theorem is complete if we can show that (10.34) implies that $\widetilde{\sigma}_\infty^* := \lim_{n\to\infty} \widetilde{\sigma}_n^* \le \sigma^*$. To effectuate this, suppose to the contrary, that there exists a $\sigma^* < \sigma < \widetilde{\sigma}_\infty^*$. Choose N so large that $4\widetilde{\lambda}_1^{(N)} < \sigma^2$ and thereafter $0 < \delta < \ell$ so small that $\Lambda_1^{(N)}(\delta) < \sigma^2/4$, where $\Lambda_1^{(N)}$ is defined by (10.17) with $\widetilde{c}_N(u)\widetilde{a}_N'(u)$ in lieu of $c(u)a'(u)$. Following the proof of Lemma 6.6(iv), equation (10.35) has a maximal solution θ_n on $[0, \delta]$ such that (10.24) holds with θ_n and $\Lambda_1^{(N)}$ in the place of $\theta(\cdot, \sigma)$ and Λ_1 respectively for every $n \ge N$. Moreover, if $[0, \delta_n]$ denotes the maximal interval of existence of θ_n contained in $[0, \ell]$ such that θ_n is positive on $(0, \delta_n)$, $\delta_{n+1} \ge \delta_n$, and, $\theta_{n+1}(s) \ge \theta_n(s)$ for all $0 \le s \le \delta_n$ and $n \ge N$. So, we can define $\delta_\infty := \lim_{n\to\infty} \delta_n$, and, $\theta_\infty(s) := \lim_{n\to\infty} \theta_n(s)$ for $0 \le s < \delta_\infty$. Letting $n \to \infty$ in (10.35) we subsequently deduce that θ_∞ solves (10.3) on $[0, \delta_\infty]$, and, (10.24) holds with θ_∞ and $\Lambda_1^{(N)}$ in the roles of $\theta(\cdot, \sigma)$ and Λ_1 respectively. By Lemma 6.8(i) though, this means that θ_∞ must be the maximal solution of (10.3). Thus, since (10.3) has a unique solution θ on $[0, \ell]$ which is positive on $(0, \ell)$ and such that $\theta(\ell) = 0$, there holds $\delta_\infty = \ell$. On the other hand, utilizing the existence of $\Theta(s) := \theta(\ell - s)$, the equation (10.16) with $c(\ell - r)a'(\ell - r)$ replaced by $\widetilde{c_N}(\ell - r)\widetilde{a_N}'(\ell - r)$ has a unique solution $\Theta_N \ge \Theta$ on $[0, \ell]$ by Lemma 2.18. Moreover, since $\sigma < \widetilde{\sigma}_N^*$, necessarily $\Theta_N(\ell) > 0$. Subsequently, by the comparison argument employed in the proof of Lemma 10.9, we have $\theta_n(s) \le \Theta_N(\ell - s)$ for all $0 \le s \le \delta_n$ and $n \ge N$. Whence, we deduce that $\theta_\infty(s) \le \Theta_N(s)$ for all $0 \le s \le \delta_\infty$. So, in particular, $\theta_\infty(\ell) = 0$. However, recalling Lemma 10.9(i), this contradicts the assumption that $\sigma > \sigma^*$. Thus the result is proved by contradiction. □

We remark that the main result of Theorem 10.17 can hardly be improved. Starting from a pair of coefficients a, c and any value $\mu > \sigma^*$, by modification of ca' in a neighbourhood of 0, it easy to construct a sequence of coefficients $\{a_n, c_n\}_{n=1}^{\infty}$ such that the corresponding parameter $\lambda_0^{(n)} = \mu^2/4$ for all $n \geq 1$, while $c_n a_n'$ converges to ca' in $L^\infty(0,\ell)$ as $n \to \infty$. Whence by Theorem 10.8 we have $\sigma_n^* \geq \mu > \sigma^*$ for all n.

Theorem 10.18. *Suppose that the hypotheses of Theorem* 10.5(b) *hold.*

(a) *Let $\mathcal{S}$ denote the set of nonincreasing, nonnegative, absolutely continuous functions on $(0,\ell)$ such that*

$$\int_0^\ell \psi(s)\,ds = 1. \tag{10.37}$$

For $\psi \in \mathcal{S}$ define the functional

$$\mathcal{F}(\psi) := 2\int_0^\ell \{c(s)a'(s)\psi(s)\,|\psi'(s)|\}^{1/2}\,ds.$$

(b) *Let $\mathcal{S}$ denote the set of nonincreasing, nonnegative, absolutely continuous functions ψ on $(0,\ell)$ such that $\psi'(s) < 0$ for almost all $0 < s < \ell$, $s^2\psi(s) \to 0$ as $s \downarrow 0$, and,*

$$\int_0^\ell \frac{\psi^2(s)}{|\psi'(s)|}\,ds = 1. \tag{10.38}$$

For $\psi \in \mathcal{S}$ define the functional

$$\mathcal{F}(\psi) := \left\{2\int_0^\ell c(s)a'(s)\psi(s)\,ds\right\}^{1/2}.$$

Then in both cases, $\sigma^* = \sup\{\mathcal{F}(\psi) : \psi \in \mathcal{S}\}$. In particular, if $\sigma^* > 2\sqrt{\lambda_1}$ there exists a $\psi \in \mathcal{S}$ such that $\mathcal{F}(\psi) = \sigma^*$.

Proof. Fix $\sigma = \sigma^*$. By Theorem 2.10 and Lemma 2.24, equation (10.3) admits a solution θ on $[0,\ell]$ which is positive on $(0,\ell)$ and satisfies $\theta(\ell) = 0$.

(a) Multiplying (10.4) by $\psi \in \mathcal{S}$ there holds

$$(\theta\psi)' = \theta\psi' + \sigma\psi - \frac{ca'\psi}{\theta} \tag{10.39}$$

almost everywhere in $(0,\ell)$. Simultaneously, it is easily verified that $s\psi(s) \to 0$ as $s \downarrow 0$ for any function $\psi \in \mathcal{S}$. Thus, invoking Lemma 6.6 there holds $\theta(s)\psi(s) \to 0$ as $s \downarrow 0$ and $s \uparrow \ell$. Integrating (10.39) from 0 to ℓ and recalling (10.37) then yields

$$\sigma = -\int_0^\ell \left\{\theta(s)\psi'(s) - \frac{c(s)a'(s)\psi(s)}{\theta(s)}\right\}\,ds,$$

or, rearranging,

$$\sigma = \mathcal{F}(\psi) + \int_0^\ell \frac{\left\{\theta(s)\sqrt{|\psi'(s)|} - \sqrt{c(s)a'(s)\psi(s)}\right\}^2}{\theta(s)}\, ds. \tag{10.40}$$

This gives $\sigma \geq \sup\{\mathcal{F}(\psi) : \psi \in \mathcal{S}\}$. Moreover, it implies that $\sigma = \mathcal{F}(\psi)$ for a $\psi \in \mathcal{S}$ whenever the last integrand in (10.40) vanishes. Elementary analysis shows that this occurs if and only if

$$\psi(s) = A\theta(s) \exp\left\{\int_s^{\ell/2} \frac{\sigma}{\theta(r)}\, dr\right\} \tag{10.41}$$

for some number $A > 0$. Now, if $\sigma > 2\sqrt{\lambda_1}$, θ must be the maximal solution of (10.3) by Lemma 10.9(ii). Whence, by Lemma 6.6(iv), (10.29) holds with σ in lieu of $\widehat{\sigma}$. In which case, it can be ascertained that the function ψ given by (10.41) is integrable on $(0, \ell)$. Consequently, for a suitable choice of A this function ψ belongs to $\mathcal{S}$, and, we have $\sigma = \mathcal{F}(\psi)$. This proves the theorem in the case $\sigma > 2\sqrt{\lambda_1}$. When $\sigma \leq 2\sqrt{\lambda_1}$, let us fix $0 < \mu < \sigma^2/4\lambda_1$, define $c_n(s) := \mu c(s)$ for $0 \leq s < \ell/n$ and $c_n(s) := c(s)$ otherwise, and, let σ_n denote the critical wave speed σ^* associated with equation (10.1) with c replaced by c_n for $n \geq 1$. By the previous theorem, $\sigma_n \to \sigma$ as $n \to \infty$. On the other hand, by what we have already proved of the present theorem, there exists a function $\psi_n \in \mathcal{S}$ such that

$$\sigma_n = 2\int_0^\ell \{c_n(s)a'(s)\psi_n(s)\,|\psi_n'(s)|\}^{1/2}\, ds \leq \mathcal{F}(\psi_n)$$

for every $n \geq 1$ such that $\sigma_n > 2\sqrt{\mu\lambda_1}$. Thus $\sigma \leq \limsup_{n\to\infty} \mathcal{F}(\psi_n) \leq \sup\{\mathcal{F}(\psi) : \psi \in \mathcal{S}\}$. Whereupon, the theorem is proved in this case too.

(b) The proof of this part is similar to that of the previous one. Instead of multiplying (10.4) by ψ we multiply this identity by $2\theta\psi$ with $\psi \in \mathcal{S}$. This gives

$$(\theta^2\psi)' = \theta^2\psi' + 2\sigma\theta\psi - 2ca'\psi$$

in lieu of (10.39). Subsequently integrating from 0 to ℓ and using (10.38) and the other stated properties of $\psi \in \mathcal{S}$, we deduce

$$\sigma^2 = \mathcal{F}^2(\psi) + \int_0^\ell \frac{\{\theta(s)\psi'(s) + \sigma\psi(s)\}^2}{|\psi'(s)|}\, ds. \tag{10.42}$$

This implies $\sigma \geq \sup\{\mathcal{F}(\psi) : \psi \in \mathcal{S}\}$, and, in addition that $\sigma = \mathcal{F}(\psi)$ for some $\psi \in \mathcal{S}$ if we can find such a function for which the last integrand in (10.42) vanishes. Here, analysis shows that this occurs if and only if

$$\psi(s) = A \exp\left\{\int_s^{\ell/2} \frac{\sigma}{\theta(r)}\, dr\right\}$$

for some number $A > 0$. In which case, by the argument employed in the proof of part (a), when $\sigma > 2\sqrt{\lambda_1}$ an A can be chosen so that such a function lies in $\mathcal{S}$. The extension to the case $\sigma \le 2\sqrt{\lambda_1}$ follows analogously. We omit further details. □

10.1.2 Wavefront properties

In 1958, Oleinik, Kalashnikov and Chzhou published their seminal paper [223] on the nonlinear degenerate diffusion equation $u_t = (a(u))_{xx}$, with the Cauchy problem for the porous media equation for $m > 1$ as prototype. Under suitable assumptions on the coefficient a and the initial data, they established the existence and uniqueness of an appropriately-defined weak solution. Furthermore, they proved that such a solution would exhibit *finite speed of propagation*. This is to say, that given initial data with compact support, the solution would continue to have compact support with respect to the spatial variable x at all later times. This property is in stark contrast to that exhibited by solutions of the heat equation. Given an initial data function which is nonnegative but not identically zero, the corresponding solution of the linear heat equation will always be positive everywhere. As a matter of interest in the framework of the present monograph, expanding on work of Barenblatt and Vishik [26], Oleinik et al. proved the occurrence of finite speed of propagation using travelling-wave solutions of the equation. It was probably Gurtin and MacCamy [140] who, precisely due to this property of finite speed of propagation, first recognized the potential offered by degenerate diffusion equations for modelling biological population migration, although such an equation was proposed earlier as a model of biological population dispersal by Gurney and Nisbet [138]. Today, degenerate reaction-diffusion equations of the type (10.1) are an accepted tool in mathematical biology [211].

The above naturally leads to the question of whether or not wavefront solutions of degenerate nonlinear reaction-diffusion equations reflect the property of finite speed of propagation. To be specific, given a wavefront solution f of equation (10.1) satisfying $f(\xi) \to 0$ as $\xi \to \infty$ and $f(\xi) \to \ell$ as $\xi \to -\infty$, is it true that

$$f(\xi) = 0 \quad \text{for all } \xi \ge \xi^* \quad \text{some } -\infty < \xi^* < \infty, \tag{10.43}$$

or, is

$$f(\xi) > 0 \quad \text{for all } -\infty < \xi < \infty? \tag{10.44}$$

Of course, one may equally as well ask, is

$$f(\xi) = \ell \quad \text{for all } \xi \le \xi^{**} \quad \text{some } -\infty < \xi^{**} < \infty, \tag{10.45}$$

or, is

$$f(\xi) < \ell \quad \text{for all } -\infty < \xi < \infty? \tag{10.46}$$

Application 2.21 with $k \ge 0$ shows that both (10.43) and (10.45) may occur. These questions have received considerable interest in recent years. Terms which have been used to designate wavefront solutions satisfying (10.43), include "weak" (*sic*) [135], *finite* [18,161,226,228], and, of *sharp type* [254–259]. Oppositely to the last two terms, wavefront solutions for which (10.44) holds have been referred to as *positive* [226], and, of *front type* [254,256,258,259], respectively.

Our next theorem addresses these questions using the integral equation approach. Rather than presenting the most general result possible, we present a conclusion which succinctly covers previous work. We use the next two lemmata. The reader so inclined may use these lemmata to formulate a more general result. Even then, it should be noted that the lemmata are not stated in the most general form possible. Details which can be used for constructing alternatives can be found in Chapter 6, cf. Theorems 6.2 and 6.9.

Lemma 10.19. *Suppose that* $c(s) > 0$ *for all* $0 < s < \delta$ *and some* $0 < \delta < \ell$, *and, that* $\lambda_0 = \lambda_1$.

(i) *If* $\sigma > 2\sqrt{\lambda_1}$, *or, if* $\sigma = 2\sqrt{\lambda_1}$ *and* $c(s)a'(s)/s \to \lambda_1$ *as* $s \downarrow 0$, *then the maximal solution* θ *of equation* (10.3) *satisfies*

$$\frac{\theta(s)}{s} \to \frac{\sigma + \sqrt{\sigma^2 - 4\lambda_1}}{2} \qquad \text{as } s \downarrow 0.$$

(ii) *If* $\sigma \geq 2\sqrt{\lambda_1}$, *and,* $c(s)a'(s)/s \to \lambda_1$ *as* $s \downarrow 0$, *then any solution* θ *of* (10.3) *other than the maximal solution satisfies*

$$\frac{\theta(s)}{s} \to \frac{\sigma - \sqrt{\sigma^2 - 4\lambda_1}}{2} \qquad \text{as } s \downarrow 0.$$

(iii) *If* $\sigma \geq 2\sqrt{\lambda_1}$, ca' *is differentiable on* $[0, \delta)$, $(ca')(0) = 0$, *and, there holds* $(ca')'(s) \to \lambda_1$ *as* $s \downarrow 0$, *then any solution* θ *of* (10.3) *other than the maximal solution satisfies*

$$\frac{\theta(s)}{c(s)a'(s)} \to \frac{2}{\sigma + \sqrt{\sigma^2 - 4\lambda_1}} \qquad \text{as } s \downarrow 0.$$

Lemma 10.20. *Suppose that* $c(s) \leq 0$ *for all* $0 < s < \delta$ *and some* $0 < \delta < \ell$. *Define*

$$Q(s) := \left| 2\int_0^s c(r)a'(r)\,dr \right|^{1/2}.$$

Suppose furthermore that (10.3) *has a unique solution* θ *on* $[0, \delta)$.

(i) *If* $\sigma = 0$, *or, if* $Q(s)/s \to \infty$ *as* $s \downarrow 0$, *then*

$$\frac{\theta(s)}{Q(s)} \to 1 \qquad \text{as } s \downarrow 0.$$

(ii) *If* $Q(s)/s \to \mu$ *as* $s \downarrow 0$ *for some* $0 \le \mu < \infty$, *then*

$$\frac{\theta(s)}{s} \to \frac{\sigma + \sqrt{\sigma^2 + 4\mu^2}}{2} \qquad \text{as } s \downarrow 0.$$

(iii) *If* $c(u) < 0$ *for all* $0 < u < \delta$, ca' *is differentiable on* $(0, \delta)$, $(ca')(0) = 0$, $(ca')'(s) \to -\mu^2$ *as* $s \downarrow 0$ *for some* $0 \le \mu < \infty$, *and,* $\sqrt{\sigma^2 + 4\mu^2} > \sigma$, *then*

$$\frac{\theta(s)}{|c(s)a'(s)|} \to \frac{2}{\sqrt{\sigma^2 + 4\mu^2} - \sigma} \qquad \text{as } s \downarrow 0.$$

Lemma 10.19 follows from Lemmata 6.6 and 6.8. Similarly, if Q is positive on $(0, \delta)$, Lemma 10.20 is a corollary of Lemma 6.3. On the other hand, if Q is not positive on $(0, \delta)$, then necessarily $ca' = 0$ almost everywhere in an interval $(0, \delta^*)$ with $0 < \delta^* < \delta$. Subsequently, the 'solution' of equation (10.3) is $\theta(s) = \sigma s$ for $0 \le s \le \delta^*$ and Lemma 10.20 is easily obtained.

Lemmata 10.19 and 10.20 complete the proof of Application 10.7 when they are applied in combination with Theorem 3.4 to the theory of Chapter 2.

The following is the promised result on finite speed of propagation. It may be viewed as a highlight of this chapter.

Theorem 10.21 (Bounded support). *Suppose that* $\ell < \infty$, $c(u) > 0$ *for all* $0 < u < \ell$, *and the parameter defined* λ_1 *by* (10.15) *is finite. Let* $\sigma^* > 0$ *denote the critical value such that equation* (10.1) *has a distinct wavefront solution from* ℓ *to* 0 *for every wave speed* $\sigma \ge \sigma^*$ *and no such solution for any wave speed* $\sigma < \sigma^*$.

(i) *Suppose furthermore that* ca' *is differentiable on* $[0, \delta]$ *for some* $0 < \delta < \ell$, $(ca')(0) = 0$, *and,* $(ca')'(u) \to (ca')'(0)$ *as* $u \downarrow 0$. *Then the following alternatives are mutually exclusive.*

(a) *Every wavefront solution from* ℓ *to* 0 *satisfies* (10.43). *This occurs if and only if*

$$\int_0^\delta \frac{1}{c(s)}\, ds < \infty.$$

(b) *Every wavefront solution from* ℓ *to* 0 *with wave speed* σ^* *satisfies* (10.43), *whereas, every such solution with wave speed* $\sigma > \sigma^*$ *satisfies* (10.44). *This occurs if and only if*

$$\int_0^\delta \frac{a'(s)}{s}\, ds < \infty = \int_0^\delta \frac{1}{c(s)}\, ds.$$

(c) *Every wavefront solution from* ℓ *to* 0 *satisfies* (10.44). *This occurs if and only if*

$$\int_0^\delta \frac{a'(s)}{s}\, ds = \infty.$$

(ii) *Suppose furthermore that ca' is differentiable on $[\ell-\delta,\ell]$ for some $0<\delta<\ell$, $(ca')(\ell)=0$, and, $(ca')'(u)\to(ca')'(\ell)$ as $u\uparrow\ell$. Then the following alternatives are mutually exclusive.*

(a) *Every wavefront solution from ℓ to 0 satisfies* (10.45). *This occurs if and only if*

$$\int_{\ell-\delta}^{\ell}\frac{1}{c(s)}\,ds<\infty.$$

(b) *Every wavefront solution from ℓ to 0 satisfies* (10.46). *This occurs if and only if*

$$\int_{\ell-\delta}^{\ell}\frac{1}{c(s)}\,ds=\infty.$$

Proof. For $\sigma\geq\sigma^*$ let θ denote the unique solution of equation (10.3) on $[0,\ell]$ satisfying $\theta(\ell)=0$. If $\sigma=\sigma^*>2\sqrt{(ca')'(0)}$, θ must be the maximal solution of the equation by Lemma 10.9(ii). Hence, by Lemma 10.19(i),

$$\theta(s)\sim\frac{\sigma+\sqrt{\sigma^2-4(ca')'(0)}}{2}s\quad\text{ as } s\downarrow 0;\tag{10.47}$$

while, if $\sigma=\sigma^*=2\sqrt{(ca')'(0)}$, irrespective of whether θ is the maximal solution of equation (10.3) or not, (10.47) holds by Lemma 10.19 parts (i) and (ii). Meanwhile, if $\sigma>\sigma^*$, θ cannot be the maximal solution of the equation, by Lemma 10.9(i). So, by Lemma 10.19(iii),

$$\theta(s)\sim\frac{2}{\sigma+\sqrt{\sigma^2-4(ca')'(0)}}c(s)a'(s)\quad\text{ as } s\downarrow 0.$$

This gives part (i) of the theorem via Corollary 2.34. To establish part (ii), we draw on the observation that $\Theta(s):=\theta(\ell-s)$ must be the unique solution of equation (10.16) on $[0,\ell]$. See, for instance, the proof of Lemma 3.5. Applying Lemma 10.20(iii) to equation (10.16) yields

$$\theta(s)\sim\frac{2}{\sigma+\sqrt{\sigma^2-4(ca')'(\ell)}}c(s)a'(s)\quad\text{ as } s\uparrow\ell.$$

Corollary 2.34 then confirms the assertions. □

The proof of Theorem 10.21 also readily yields the next result, recalling how wavefront solutions can be constructed from solutions of the integral equation as described in Chapter 2.

Theorem 10.22 (Asymptotic behaviour). *Let f denote a wavefront solution of equation* (10.1) *from ℓ to 0 with wave speed σ.*

(i) *Suppose that the conditions of Theorem* 10.21 *part* (i) *hold. Define*

$$\Xi_0 := \sup\{\xi \in (-\infty, \infty) : f(\xi) > 0\}. \tag{10.48}$$

Then, if $\sigma = \sigma^*$ *there holds*

$$\frac{(a(f))'(\xi)}{f(\xi)} \to -\frac{\sigma + \sqrt{\sigma^2 - 4(ca')'(0)}}{2} \qquad \text{as } \xi \uparrow \Xi_0,$$

whereas, if $\sigma > \sigma^*$ *there holds*

$$\frac{f'(\xi)}{c(f(\xi))} \to -\frac{2}{\sigma + \sqrt{\sigma^2 - 4(ca')'(0)}} \qquad \text{as } \xi \uparrow \Xi_0.$$

(ii) *Suppose that the conditions of Theorem* 10.21 *part* (ii) *hold. Define*

$$\Xi_1 := \inf\{\xi \in (-\infty, \infty) : f(\xi) < \ell\}. \tag{10.49}$$

Then

$$\frac{f'(\xi)}{c(f(\xi))} \to -\frac{2}{\sigma + \sqrt{\sigma^2 - 4(ca')'(\ell)}} \qquad \text{as } \xi \downarrow \Xi_1.$$

For the semilinear equation $u_t = u_{xx} + c(u)$ with $c \in C^1([0,1])$, $c(0) = c(1) = 0$ and $c(u) > 0$ for $0 < u < 1$, it follows from Lemma 2.26 and Theorem 10.21 that any wavefront solution f from 1 to 0 is necessarily strictly decreasing in $(-\infty, \infty)$.

It may of significance that as a supplement to the results in Theorem 10.22, estimates for the solution of the integral equation (10.3) on the whole of $[0, \ell]$ may be used to obtain global estimates of wavefront solutions.

Theorem 10.23. *Suppose that* ca' *is continuous in* $[0, \ell]$ *and differentiable in* $(0, \ell)$, $(ca')(0) = (ca')(\ell) = 0$ *and* $c(u) > 0$ *for* $0 < u < \ell$. *Set* $A := \inf\{(ca')'(u) : 0 < u < \ell\}$ *and* $B := \sup\{(ca')'(u) : 0 < u < \ell\}$. *Then any wavefront solution* f *of* (10.1) *with wave speed* σ *from* ℓ *to* 0 *satisfies*

$$\frac{f'(\xi)}{c(f(\xi))} \le -\frac{2}{\sigma + \sqrt{\sigma^2 - 4A}} \qquad \text{for all } -\infty < \xi < \infty,$$

and for any such wavefront solution with wave speed $\sigma \ge 2\sqrt{B}$ *there holds*

$$\frac{f'(\xi)}{c(f(\xi))} \ge -\frac{2}{\sigma + \sqrt{\sigma^2 - 4B}} \qquad \text{for all } -\infty < \xi < \infty.$$

Proof. Note that for this example $A < 0 \le \lambda_0 = \lambda_1 \le \Lambda_1(\ell) \le B$. The results follow from the observation that setting $\beta_1 := 2/(\sigma + \sqrt{\sigma^2 - 4A})$ and $\beta_2 := 2/(\sigma + \sqrt{\sigma^2 - 4B})$ when $\sigma \ge 2\sqrt{B}$, the functions $\Theta_i(s) := \beta_i c(\ell - s)a'(\ell - s)$ are solutions of equation (10.16) on $[0, \ell]$ with $-\sigma s$ replaced by $\Theta_i(s) - s/\beta_i$ for $i = 1, 2$

respectively. Moreover, $s \mapsto -\sigma s - \Theta_1(s) + s/\beta_1$ is an increasing function on $[0,\ell]$, and, $s \mapsto -\sigma s - \Theta_2(s) + s/\beta_2$ is a decreasing function on $[0,\ell]$ for $\sigma \geq 2\sqrt{B}$. Consequently Lemmata 2.14 and 2.18 imply that the unique solution Θ of (10.16) on $[0,\ell]$ is such that $\Theta \geq \Theta_1$ on $[0,\ell]$, and, such that $\Theta \leq \Theta_2$ on $s \in [0,\ell]$ if $\sigma \geq 2\sqrt{B}$. Recalling how f is defined in terms of the solution θ of (10.3) satisfying the integrability condition on $[0,\ell]$ with $\theta(\ell) = 0$ and how this function θ is defined in terms of Θ yields the desired estimates. □

A corollary of this theorem is that for large σ one has the estimates

$$-B + \mathcal{O}(\sigma^{-2}) \leq \sigma^3 \frac{f'(\xi)}{c(f(\xi))} + \sigma^2 \leq -A + \mathcal{O}(\sigma^{-2})$$

for $-\infty < \xi < \infty$. This led Kelley [180] to search for and obtain estimates of the form

$$-\mathcal{B}\sigma^{-2} \leq \sigma^3 \frac{f'(\xi)}{c(f(\xi))} + \sigma^2 + (ca')'(f(\xi)) \leq -\mathcal{A}\sigma^{-2},$$

where

$$\mathcal{A} < \inf\{\left(2\{(ca')'\}^2 + ca'(ca')''\right)(s) : 0 < s < \ell\}$$

and

$$\mathcal{B} > \sup\{\left(2\{(ca')'\}^2 + ca'(ca')''\right)(s) : 0 < s < \ell\},$$

for large enough σ. Under the additional assumptions that $(ca')'$ is bounded and differentiable in $(0,\ell)$ and $ca'(ca')''$ is bounded, these estimates can also be found by applying the technique used above to estimate solutions of the integral equation (8.1). The appropriate test functions are

$$\Theta_i(s) := \sigma^{-3}\left\{\sigma^2 + (ca')'(\ell - s) + \sigma^{-2}\beta_i\right\} c(\ell - s)a'(\ell - s)$$

where $\beta_1 = \mathcal{A}$ and $\beta_2 = \mathcal{B}$.

We close this section by considering two further examples of reaction-diffusion equations which have attracted especial interest in the past. The first is Murray's model of biological dispersal in which the unknown u represents population density [211]. This is the equation

$$u_t = (u^{m-1}u_x)_x + u^p(1 - u^q) \tag{10.50}$$

in which m, p and q are positive parameters. This model includes the Fisher equation as a special case. Travelling-wave solutions of this equation have been investigated extensively in the past.

Application 10.24. *Suppose that m, p and q are positive constants.*

(i) *If $m+p < 2$, then equation* (10.50) *does not admit a wavefront solution from* 1 *to* 0.

(ii) *If $m+p \geq 2$, then there exists a $\sigma^* > 0$, which depends only on $m+p$ and q, such that equation (10.50) admits precisely one distinct wavefront solution from 1 to 0 for every wave speed $\sigma \geq \sigma^*$ and no such solution for any wave speed $\sigma < \sigma^*$. The critical wave speed σ^* is a continuous function of $m+p$ and q, for fixed q is a strictly decreasing function of $m+p$, for fixed $m+p$ is a strictly increasing function of q, and, satisfies*

$$\sigma^* = 2 \quad \text{if } m+p=2,$$

$$\sigma^* > \frac{1}{\sqrt{m+p-1}} \qquad \text{if } m+p < q+2, \tag{10.51}$$

$$\sigma^* = \frac{1}{\sqrt{q+1}} \qquad \text{if } m+p = q+2, \tag{10.52}$$

$$\sigma^* < \frac{1}{\sqrt{m+p-1}} \qquad \text{if } m+p > q+2, \tag{10.53}$$

and,

$$\sqrt{2\Lambda_1(1)} < \sigma^* < \min\{2\sqrt{\Lambda_1(1)}, \sqrt{2\Lambda_2}\} \quad \text{if } m+p>2, \tag{10.54}$$

where

$$\Lambda_1(1) := \frac{q\{(m+p-2)(m+p+q-1)\}^{(m+p-2)/q}}{\{(m+p-1)(m+p+q-2)\}^{(m+p+q-2)/q}}$$

and

$$\Lambda_2 := \frac{q}{(m+p-2)(m+p+q-2)}.$$

(a) *If $p < 1$, then the support of every wavefront solution f from 1 to 0 is bounded above, and, defining Ξ_0 by (10.48), there holds*

$$f^{m-2}(\xi)f'(\xi) \to -\frac{2}{\sigma+\sqrt{\sigma^2-4}} \qquad \text{if } m+p=2, \tag{10.55}$$

$$f^{m-2}(\xi)f'(\xi) \to -\sigma \quad \text{if } m+p>2 \text{ and } \sigma=\sigma^*, \tag{10.56}$$

and,

$$f^{-p}(\xi)f'(\xi) \to -\frac{1}{\sigma} \quad \text{if } m+p>2 \text{ and } \sigma>\sigma^* \tag{10.57}$$

as $\xi \uparrow \Xi_0$.

(b) *If $m > 1$ and $p \geq 1$, then the support of every wavefront solution f from 1 to 0 with wave speed σ^* is bounded above, and, defining Ξ_0 by (10.48), f satisfies (10.56) as $\xi \uparrow \Xi_0$. Whereas, the support of every such solution f with wave speed $\sigma > \sigma^*$ is not bounded above, and, f satisfies (10.57) as $\xi \to \infty$.*

(c) *If $m \le 1$, then the support of every wavefront solution f from 1 to 0 is not bounded above, and, f satisfies* (10.55) – (10.57) *as $\xi \to \infty$.*

Every wavefront solution f from 1 to 0 with wave speed σ is such that $f(\xi) < 1$ for all $-\infty < \xi < \infty$, and,

$$\{1 - f(\xi)\}^{-1} f'(\xi) \to -\frac{2q}{\sigma + \sqrt{\sigma^2 + 4q}} \qquad \text{as } \xi \to -\infty.$$

Proof. In terms of the general theory, for this example $(ca')(s) = s^{m+p-1}(1 - s^q)$. So we easily compute that $\lambda_1 = \infty$ if $m+p < 2$, $\lambda_0 = \lambda_1 = \Lambda_1(1) = 1$ if $m+p = 2$; while, $\lambda_0 = \lambda_1 = 0$, and, $\Lambda_1(1)$ and Λ_2 are as stated, if $m + p > 2$. Theorem 10.5 then gives the existence results, Theorem 10.8 the exact value of σ^* in the case $m + p = 2$, and, the estimate (10.54) in the case $m + p > 2$, Theorem 10.16 the monotonicity of σ^*, and, Theorem 10.17 the continuity of σ^*. Furthermore, noting that $ca' \in C^1([0, 1])$ if $m + p \ge 2$, Theorems 10.21 and 10.22 give the results on the behaviour of a wavefront solution for large $|\xi|$. To establish (10.51) – (10.53), we note that if $m + p < q + 2$ there holds $s^{m+p}(1 - s^q) > s^{m+p}(1 - s^{m+p-2})$, if $m + p = q + 2$ there holds $s^{m+p}(1 - s^q) = s^{m+p}(1 - s^{m+p-2})$, while, if $m + p > q + 2$ there holds $s^{m+p}(1 - s^q) < s^{m+p}(1 - s^{m+p-2})$ for all $0 < s < 1$. However, by Application 10.14, we know that the critical wave speed associated with the equation $u_t = u_{xx} + u^{m+p}(1 - u^{m+p-2})$ is $\sigma = 1/\sqrt{m + p - 1}$. The inequalities (10.51) – (10.53) subsequently follow from Theorem 10.16. □

Our final example in this section is the equation

$$u_t = (u^{m-1}(1 - u)^{n-1} u_x)_x + u(1 - u). \tag{10.58}$$

Application 10.25. *Suppose that m and n are positive constants.*

(i) *If $m < 1$, then equation* (10.58) *does not admit a wavefront solution from* 1 *to* 0.

(ii) *If $m \ge 1$, then there exists a $\sigma^* > 0$, which depends only on m and n, such that equation* (10.58) *admits precisely one distinct wavefront solution from* 1 *to* 0 *for every wave speed $\sigma \ge \sigma^*$ and no such solution for any wave speed $\sigma < \sigma^*$. The critical wave speed σ^* is a continuous function of m and n, for fixed n is a strictly decreasing function of m, for fixed m is a strictly decreasing function of n, and, satisfies $\sigma^* = 2$ if $m = 1$, and, $\sigma^* = 1/\sqrt{2}$ if $m = 2$ and $n = 1$.*

(a) *If $m = 1$, then the support of every wavefront solution f from* 1 *to* 0 *is not bounded above, and,*

$$f^{-1}(\xi) f'(\xi) \to -\frac{2}{\sigma + \sqrt{\sigma^2 - 4}} \qquad \text{as } \xi \to \infty.$$

(b) *If $m > 1$, then the support of every wavefront solution f from 1 to 0 with wave speed σ^* is bounded above, and, defining Ξ_0 by (10.48),*

$$f^{m-2}(\xi)f'(\xi) \to -\sigma \qquad \text{as } \xi \uparrow \Xi_0.$$

Whereas, the support of every such solution f with wave speed $\sigma > \sigma^*$ is not bounded above, and,

$$f^{-1}(\xi)f'(\xi) \to -\frac{1}{\sigma} \qquad \text{as } \xi \to \infty.$$

Every wavefront solution f from 1 to 0 with wave speed σ is such that $f(\xi) < 1$ for all $-\infty < \xi < \infty$, and,

$$\{1 - f(\xi)\}^{(n-3)/2} f'(\xi) \to -\sqrt{\frac{2}{n+1}} \qquad \text{if } n < 1,$$

$$\{1 - f(\xi)\}^{-1} f'(\xi) \to -\frac{2}{\sigma + \sqrt{\sigma^2 + 4}} \qquad \text{if } n = 1,$$

and,

$$\{1 - f(\xi)\}^{-1} f'(\xi) \to -\frac{1}{\sigma} \qquad \text{if } n > 1$$

as $\xi \to -\infty$.

Proof. The proof of this example runs along the same lines as that of the previous one. The only complication is that if $n < 1$ the conditions assumed in part (ii) of Theorem 10.21 are not met. For this case, we fall back on Lemma 10.20 part (i) applied to equation (10.16), where $\Theta(s) = \theta(1 - s)$, to deduce that the unique solution θ of the integral equation (10.3) on $[0, 1]$ satisfying $\theta(1) = 0$ is such that $\theta(s) \sim \sqrt{2/n+1}\,(1 - s)^{(n+1)/2}$ as $s \uparrow 1$. With this information, we can use the same strategy as in Application 10.24. We leave the details to the reader. □

In [22] the case $m = 1$ and $n = 0$ in the above example is also considered. Ostensibly this falls out of the scope of our theory, since when this equation is written in the general form (10.1) one does not have the hypothesis that a is continuous on the whole of I, but merely that a is continuous on $(0, \ell)$. This objection can be overcome by the expedient of tightening the definition of a wavefront solution. If $a(u)$ becomes unbounded as $u \downarrow 0$, one has to impose the additional requirements that any wavefront solution f of the equation from ℓ to 0 is such that $f(\xi) > 0$ for all $\xi \in (-\infty, \infty)$, and, that $(a(f))'(\xi) \to 0$ as $\xi \to \infty$. When $a(u)$ remains bounded as $u \downarrow 0$, the latter property is a natural consequence of the convergence of $f(\xi)$ as $\xi \to \infty$, cf. Corollary 2.31. Similarly, if $a(u)$ becomes unbounded as $u \uparrow \ell < \infty$ one has to impose the additional requirements that any wavefront solution f of the equation from ℓ to 0 is such that $f(\xi) < \ell$ for all $\xi \in (-\infty, \infty)$, and, that $(a(f))'(\xi) \to 0$ as $\xi \to -\infty$. With these adaptations, it can be seen that our proof of Theorem 2.10 remains valid. In which case, the above example extends to the case $m = 1$ and $n = 0$. Indeed, with the adaptations suggested, the conclusions of the example apply to any $m > -1$ and $n > -1$.

10.2 One sign change

The previous section has been concerned with nonlinear reaction-diffusion equations of the type (10.1) where the reaction term is strictly positive on $(0,\ell)$. By implication, through Theorem 3.4, this also covers the case where the reaction term is strictly negative on $(0,\ell)$. We turn now to the case that the reaction term has one change of sign on $(0,\ell)$. By this we mean that there is a $0<\alpha<\ell$ such that $(\alpha-u)c(u)\geq 0$ for all $0<u<\ell$, or, $(\alpha-u)c(u)\leq 0$ for all $0<u<\ell$. Notwithstanding, in the first of these two cases, it can be determined that an equation of the form (10.1) cannot possibly admit a wavefront solution from ℓ to 0, as a consequence of Theorem 10.1. Therefore, we shall only pursue the second case. We call α the point of sign change.

Since we shall use the following notation repeatedly in the rest of this section, let us forthwith define

$$Q_0(s) := \left|2\int_0^s c(r)a'(r)\,dr\right|^{1/2} \tag{10.59}$$

and

$$Q_1(s) := \left|2\int_s^\ell c(r)a'(r)\,dr\right|^{1/2}. \tag{10.60}$$

Our first result in this section concerns sufficient criteria for the existence of a wavefront solution. Complementary necessary criteria are provided by Theorem 10.1.

Theorem 10.26 (Existence). *Suppose that $c(u)\leq 0$ for all $0<u<\alpha$, and, $c(u)\geq 0$ for all $\alpha<u<\ell$, for some $0<\alpha<\ell<\infty$. Let κ be defined by (10.2). Suppose furthermore that one of the following holds:*

(a) *$\kappa>0$, and, $c(u)>0$ for all $\alpha<u<\ell$;*

(b) *$\kappa=0$, $Q_0(u)>0$ for all $0<u<\alpha$, and, $Q_1(u)>0$ for all $\alpha<u<\ell$;*

(c) *$\kappa<0$, and, $c(u)<0$ for all $0<u<\alpha$.*

Then there exists a real number σ^* such that equation (10.1) has exactly one distinct wavefront solution from ℓ to 0 with wave speed σ^* and no such wavefront solution with any other wave speed.

Proof. Invoking Lemma 2.14(i), let $\mathcal{S}_0$ denote the set of values σ such that (10.3) has a unique solution $\theta(s;\sigma)$ on $[0,\alpha]$, $\mathcal{S}_1$ denote the set of values σ such that (10.16) has a unique solution $\Theta(s;\sigma)$ on $[0,\ell-\alpha]$, and, $\mathcal{S}:=\mathcal{S}_0\cap\mathcal{S}_1$. By substitution, it is easily verified that Q_0 solves (10.3) with $\sigma=0$ on $[0,\alpha]$, while $Q_1(\ell-s)$ solves (10.16) with $\sigma=0$ on $[0,\ell-\alpha]$. So, by Lemma 8.5, $\mathcal{S}_0$ is an interval containing $[0,\infty)$ with $\theta(s;0)=Q_0(s)$, and, $\mathcal{S}_1$ is an interval containing $(-\infty,0]$ with $\Theta(s;0)=Q_1(\ell-s)$. Furthermore, using equation (10.3) it can be determined that $\theta(\alpha;\sigma)\to\infty$ as $\sigma\to\infty$, and, similarly using (10.16) that $\Theta(\ell-\alpha;\sigma)\to\infty$

as $\sigma \to -\infty$, cf. [124]. For $\sigma \in \mathcal{S}$, we define $F(\sigma) := \theta(\alpha;\sigma) - \Theta(\ell-\alpha;\sigma)$. By Lemma 8.5, F is a continuous function of $\sigma \in \mathcal{S}$; while,

$$F(0) = Q_0(\ell) - Q_1(\alpha) = -\frac{2\kappa}{Q_0(\alpha)+Q_1(\ell)}.$$

Recalling Theorem 2.10 and Lemma 8.9, to prove that the existence of a wave speed σ^* for which equation (10.1) has a wavefront solution from ℓ to 0, it suffices to show that there is a $\sigma^* \in \mathcal{S}$ for which $F(\sigma^*) = 0$, $\theta(\cdot;\sigma^*) > 0$ on $(0,\alpha]$, and, $\Theta(\cdot;\sigma^*) > 0$ on $(0,\ell-\alpha]$. For this purpose, we distinguish the three cases in the statement of the theorem.

(a) In this case $\mathcal{S}_1 = (-\infty,\infty)$ by Lemma 2.16 part (i) applied to equation (10.16). Hence, $\mathcal{S}$ contains $[0,\infty)$. Furthermore, we have $F(0) < 0$ and $F(\sigma) \to \infty$ as $\sigma \to \infty$. The continuity of F subsequently infers the existence of a $\sigma^* > 0$ such that $F(\sigma^*) = 0$. Moreover, since by (10.3) there holds $\theta(s;\sigma^*) \geq \sigma^* s$ for all $0 < s \leq \alpha$, and, by Lemma 2.16(i) applied to (10.16) there holds $\Theta(s;\sigma^*) > 0$ for all $0 < s < \ell-\alpha$, the functions $\theta(\cdot;\sigma^*)$ and $\Theta(\cdot;\sigma^*)$ possess the required properties.

(b) In this case the hypotheses are such that $\sigma^* = 0$ fulfils the criterion.

(c) This case is the 'mirror image' of the first one. The set $\mathcal{S}$ contains $(-\infty,0]$, $F(0) > 0$, and, $F(\sigma) \to -\infty$ as $\sigma \to -\infty$. We omit further details.

Now, since $\theta(\alpha;\sigma^*) = \Theta(\ell-\alpha;\sigma^*) > 0$, by application of Lemma 2.18(i) to equations (10.3) and (10.16) there holds $F(\sigma) < 0$ for all $\sigma \in \mathcal{S} \cap (-\infty,\sigma^*)$, and, $F(\sigma) > 0$ for all $\sigma \in \mathcal{S} \cap (\sigma^*,\infty)$. So, by Theorem 2.10 and Lemma 8.9, σ^* is the only wave speed for which equation (10.1) can admit a wavefront solution from ℓ to 0. The distinctness of the wavefront solution finally follows Lemma 2.26. □

Corollary 10.27 (of Theorem 10.26). *Suppose that $c(u) < 0$ for all $0 < u < \alpha$, and, $c(u) > 0$ for all $\alpha < u < \ell$, for some $0 < \alpha < \ell < \infty$. Then there exists a real number σ^* such that equation (10.1) has exactly one distinct wavefront solution from ℓ to 0 with wave speed σ^* and no wavefront solution from ℓ to 0 for any other wave speed.*

Applications 10.2 – 10.4 are illustrations of the above corollary. In particular, Application 10.2 with $p = 1$ and Application 10.14 with $p = 1$ and the remaining numbers chosen appropriately leads to the following result.

Application 10.28. *The Nagumo equation*

$$u_t = u_{xx} + u(1-u)(u-\alpha)$$

with $0 < \alpha < 1$ admits exactly one distinct wavefront solution from α to 0 for all wave speeds $\sigma \leq \sigma_0^*$, where

$$\sigma_0^* := \begin{cases} -2\sqrt{\alpha(1-\alpha)} & \text{for } \alpha < 2/3 \\ -(2-\alpha)/\sqrt{2} & \text{for } \alpha \geq 2/3; \end{cases}$$

exactly one distinct wavefront solution from 1 *to* α *for all wave speeds* $\sigma \geq \sigma_1^*$, *where*

$$\sigma_1^* := \begin{cases} (1+\alpha)/\sqrt{2} & \text{for } \alpha \leq 1/3 \\ 2\sqrt{\alpha(1-\alpha)} & \text{for } \alpha > 1/3; \end{cases}$$

exactly one distinct wavefront solution from 1 *to* 0 *with wave speed* $\sigma = \sigma^*$, *where*

$$\sigma^* := (1-2\alpha)/\sqrt{2};$$

and no other decreasing wavefront solutions.

As a simple confirmation of Theorem 10.26 when the reaction term does not change from being strictly negative to strictly positive, consider the equation

$$u_t = u_{xx} - \begin{cases} c_0 u & \text{for } 0 \leq u \leq \alpha \\ 0 & \text{for } \alpha < u \leq 1. \end{cases} \tag{10.61}$$

This equation arises as the model of the cooling of a hot reactor core by water [204].

Application 10.29. *Let* $c_0 > 0$ *and* $0 < \alpha < 1$ *and set*

$$\sigma^* = -\alpha\sqrt{\frac{c_0}{1-\alpha}}.$$

Then equation (10.61) *admits a wavefront solution from* 1 *to* 0 *with wave speed* σ *if and only if* $\sigma = \sigma^*$.

The verification of this result is analogous to that of Application 10.4. It is easily checked that when $\sigma \neq \sigma^*$ equation (10.3) does not admit a solution θ on $[0,1]$ for which $\theta(1) = 0$. On the other hand, when $\sigma = \sigma^*$ it admits a unique solution given by $\sqrt{c_0(1-\alpha)}s$ for $0 \leq s \leq \alpha$ and $\alpha\sqrt{c_0/(1-\alpha)}(1-s)$ for $\alpha < s \leq 1$. As in Application 10.4, the wavefront solution can actually be found explicitly,

$$f(\xi) = \begin{cases} 1-(1-\alpha)\exp\{\alpha\sqrt{c_0/(1-\alpha)}(\xi-\xi_0)\} & \text{for } \xi < \xi_0 \\ \alpha\exp\{\sqrt{c_0(1-\alpha)}(\xi_0-\xi)\} & \text{for } \xi \geq \xi_0, \end{cases}$$

where $-\infty < \xi_0 < \infty$ is chosen so that $f(\xi_0) = \alpha$.

In Applications 10.2 – 10.4 and 10.29, the unique wave speed σ^* can be computed explicitly. For the general case, we have the following estimates of its magnitude.

Theorem 10.30 (Critical speed estimates). *Suppose that the hypotheses of Theorem* 10.26 *hold. Define* Q_0 *and* Q_1 *by* (10.59) *and* (10.60) *respectively. Let* σ^* *denote the unique wave speed such that equation* (10.1) *has a wavefront solution from* ℓ *to* 0.

(a) *In case* (a) *of Theorem* 10.26 *there holds*

$$\sigma^* > \max_{\gamma \leq s \leq \ell} \frac{Q_1(\alpha) - Q_0(\alpha) - Q_1(s)}{s}, \tag{10.62}$$

$$\sigma^* < \min_{\alpha \le s \le \gamma} \frac{\sqrt{2\kappa} - Q_1(\alpha) + Q_0(\alpha) + Q_1(s) - Q_0(s)}{s}, \tag{10.63}$$

and,

$$\max_{\gamma \le s \le \ell} \frac{Q_0^2(s)}{2\int_0^s Q_1(r)\,dr} < \sigma^* < \frac{\kappa}{\int_0^\gamma Q_0(r)\,dr}, \tag{10.64}$$

where γ is the unique point in $[\alpha, \ell)$ at which Q_0 vanishes.

(b) *In case* (b) *of Theorem* 10.26 *there holds* $\sigma^* = 0$.

(c) *In case* (c) *of Theorem* 10.26 *there holds*

$$\sigma^* < \min_{0 \le s \le \gamma} \frac{Q_1(\alpha) - Q_0(\alpha) + Q_0(s)}{\ell - s},$$

$$\sigma^* > \max_{\gamma \le s \le \alpha} \frac{Q_0(\alpha) - Q_1(\alpha) - \sqrt{-2\kappa} + Q_0(s) - Q_1(s)}{\ell - s},$$

and,

$$\frac{\kappa}{\int_\gamma^\ell Q_1(r)\,dr} < \sigma^* < \min_{0 \le s \le \gamma} \frac{-Q_1^2(s)}{2\int_s^\ell Q_0(r)\,dr},$$

where γ is the unique point in $(0, \alpha]$ at which Q_1 vanishes.

Proof. We prove part (a) only. Part (b) is already covered by Theorem 10.1, while part (c) may be obtained by analogy to part (a). Let θ denote the unique solution of (10.3) with $\sigma = \sigma^*$ on $[0, \ell]$ such that $\theta(\ell) = 0$. Building on the proof of Theorem 10.26, we know that $\sigma^* > 0$, and, applying Lemmata 2.17 and 2.18(i), that $\theta(s) = \theta(s; \sigma^*) > \theta(s; 0) = Q_0(s)$ for all $0 < s \le \gamma$, $\theta(s) > 0$ for all $\gamma \le s < \ell$, and $\theta(s) = \Theta(\ell - s; \sigma^*) < \Theta(\ell - s; 0) = Q_1(s)$ for all $0 \le s < \ell$. Substituting these inequalities in (10.3) with $\sigma = \sigma^*$ there holds

$$Q_1(s) > \sigma^* s + Q_1(r)\Big|_0^\alpha + Q_0(r)\Big|_\alpha^s \quad \text{for all } \alpha \le s \le \gamma$$

and

$$0 < \sigma^* s + Q_0(r)\Big|_0^\alpha + Q_1(r)\Big|_\alpha^s \quad \text{for all } \gamma \le s \le \ell.$$

Alternatively substituting them in (10.5) with $\sigma = \sigma^*$,

$$\frac{1}{2}Q_1^2(s) > \sigma^* \int_0^s Q_0(r)\,dr + \frac{1}{2}Q_0^2(s) \quad \text{for all } 0 \le s \le \gamma$$

and

$$0 < \sigma^* \int_0^s Q_1(r)\,dr - \frac{1}{2}Q_0^2(s) \quad \text{for all } \gamma \le s \le \ell.$$

Rewriting these inequalities yields (10.62) – (10.64). □

Corollary 10.31 (of Theorem 10.30). *Suppose that $c(u) = 0$ for all $0 < u < \alpha$, and, $c(u) > 0$ for all $\alpha < u < \ell$, for some $0 < \alpha < \ell < \infty$. Let $\kappa > 0$ be given by (10.2). Then there exists a real number*

$$\frac{\sqrt{2\kappa}}{\ell} < \sigma^* < \frac{\sqrt{2\kappa}}{\alpha} \tag{10.65}$$

such that equation (10.1) has exactly one distinct wavefront solution from ℓ to 0 with wave speed σ^* and no wavefront solution from ℓ to 0 for any other wave speed.

Proof. If $c \equiv 0$ on $[0, \alpha)$, then $Q_0(\alpha) = 0$, $Q_1(\alpha) = \sqrt{2\kappa}$ and $\gamma = \alpha$ in the terms of Theorem 10.30(a). Combining Theorem 10.26, (10.62) with ℓ substituting for the optimal value of s, and, (10.63) yields the corollary. □

Corollary 10.31 was essentially deduced by Zel'dovich in his highly acclaimed paper [314] published in 1948. As pointed out in [314], in the singular limit $\kappa \not\to 0$ as $\alpha \uparrow \ell$, the upper and lower bounds in (10.65) coincide. Thus, in a certain sense, the estimates in (10.62) and (10.63) are sharp. By implication through Theorem 3.4 the same subsequently also applies to their counterparts in Theorem 10.30 part (c).

As supplements to Theorem 10.30 we state the following.

Theorem 10.32. *Consider equation (10.1) with two different sets of coefficients a_1, c_1 and a_2, c_2 such that the conditions of Theorem 10.26 hold on the same interval $[0, \ell]$. Let κ_i and σ_i^* denote the appropriate parameters associated with (10.1) for $i = 1, 2$. Suppose that $c_1(u)a_1'(u) \geq c_2(u)a_2'(u)$ for all $0 < u < \ell$, and, $\kappa_1 > \kappa_2$. Then $\sigma_1^* > \sigma_2^*$.*

Proof. From the proof of Theorem 10.26, equation (10.3) with $\sigma = \sigma_i^*$ and $c(r)a'(r)$ replaced by $c_i(r)a_i'(r)$ admits a unique solution θ_i on $[0, \ell]$ which is positive on $(0, \ell)$ and satisfies $\theta_i(\ell) = 0$. Now, if $\sigma_1^* \leq \sigma_2^*$ and $c_1 a_1' \geq c_2 a_2'$ on $(0, \ell)$, we have $\theta_1 \leq \theta_2$ on $[0, \ell]$ by Lemma 2.18(i). However, by (10.7) this means that $\kappa_1 \leq \kappa_2$. Thus, the hypotheses that $c_1 a_1' \geq c_2 a_2'$ on $(0, \ell)$ and $\kappa_1 > \kappa_2$ can lead to no other conclusion than that asserted. □

Theorem 10.33. *Consider equation (10.1) with a fixed set of coefficients and with a sequence of coefficients $\{a_n, c_n\}_{n=1}^{\infty}$ such that the conditions of Theorem 10.26 hold on the same interval $[0, \ell]$. Let σ^* and $\{\sigma_n^*\}_{n=1}^{\infty}$ denote the corresponding critical wave speeds, and, θ^* be the unique solution of equation (10.3) with $\sigma = \sigma^*$ on $[0, \ell]$. Suppose that*

$$\sup_{0<u<\ell} \frac{|c_n(u)a_n'(u) - c(u)a'(u)|}{\theta^*(u)} \to 0 \quad \text{as } n \to \infty.$$

Then $\sigma_n^* \to \sigma^*$ as $n \to \infty$.

Proof. Define the function θ_n^* for the reaction-diffusion equation with coefficients a_n, c_n by analogy to θ^* for equation (10.1) with coefficients a, c. Pick $\varepsilon > 0$ and let N be so large that $c_n(s)a_n'(s) \leq c(s)a'(s) + \varepsilon\theta^*(s)$ for all $0 < s < \ell$ and $n \geq N$.

Consider equation (10.3) with $\sigma = \sigma^* + \varepsilon$ and $c(r)a'(r)$ replaced by $c(r)a'(r) + \varepsilon\theta^*(r)$. It is easily checked that θ^* itself solves this equation on $[0,\ell]$. Hence, by Lemma 2.18(i), the integral equation (10.36) with $\sigma = \sigma^* + \varepsilon$ has a solution $\theta_n \geq \theta^*$ on $[0,\ell]$ for every $n \geq N$. Now, should there hold $\sigma_n^* > \sigma^* + \varepsilon$ for $n \geq N$, by a second application of Lemma 2.18(i) to equation (10.36), there must hold $\theta_n^* \geq \theta_n$ on $[0,\ell]$. However, this infers that θ_n is a solution of (10.36) with $\sigma = \sigma^* + \varepsilon$ on $[0,\ell]$, that θ_n is positive on $(0,\ell)$, and, $\theta_n(\ell) = 0$. So, by Theorem 2.10, the reaction-diffusion equation with coefficients a_n, c_n has a wavefront solution from ℓ to 0 with wave speed $\sigma^* + \varepsilon$. This contradicts the uniqueness of the wave speed σ_n^* proven in Theorem 10.26. We conclude that $\sigma_n^* \leq \sigma^* + \varepsilon$ for all $n \geq N$. In view of the arbitrariness of ε, this leads to the result: $\limsup_{n\to\infty} \sigma_n^* \leq \sigma^*$. The assertion that $\liminf_{n\to\infty} \sigma_n^* \geq \sigma^*$ can be proven similarly, with equation (10.16) in the role of (10.3). □

Just as in the case that the reaction term is positive or negative, when c has one sign change, one may analyse the question of whether or not a wavefront solution of an equation of the form (10.1) displays finite speed of propagation. The next theorem provides an answer. As in the previous section, we shall not strive to present the most general result.

Theorem 10.34 (Bounded support). *Let f denote a wavefront solution of equation* (10.1) *from ℓ to 0 with wave speed σ. Define κ by* (10.2), *Q_0 by* (10.59), *and, Q_1 by* (10.60).

(i) *Suppose that $c \leq 0$ on $(0,\delta]$ for some $0 < \delta < \ell/2 < \infty$, ca' is differentiable on $[0,\delta]$, $(ca')(0) = 0$, and, $(ca')'(u) \to (ca')'(0)$ as $u \downarrow 0$.*

   (a) *If $\kappa > 0$, then f satisfies* (10.43) *if and only if*

$$\int_0^\delta \frac{a'(s)}{s}\,ds < \infty.$$

   (b) *If $\kappa = 0$, and, $Q_0 > 0$ on $(0,\delta]$, then f satisfies* (10.43) *if and only if*

$$\int_0^\delta \frac{a'(s)}{Q_0(s)}\,ds < \infty.$$

   (c) *If $\kappa < 0$, and, $c < 0$ on $(0,\delta]$, then f satisfies* (10.43) *if and only if*

$$\int_0^\delta \frac{1}{|c(s)|}\,ds < \infty.$$

(ii) *Suppose that $c \geq 0$ on $[\ell-\delta,\ell)$ for some $0 < \delta < \ell/2 < \infty$, ca' is differentiable on $[\ell-\delta,\ell]$, $(ca')(\ell) = 0$, and, $(ca')'(u) \to (ca')'(\ell)$ as $u \uparrow \ell$.*

   (a) *If $\kappa > 0$, and, $c > 0$ on $[\ell-\delta,\ell)$, then f satisfies* (10.45) *if and only if*

$$\int_{\ell-\delta}^{\ell} \frac{1}{c(s)}\,ds < \infty.$$

(b) *If* $\kappa = 0$, *and,* $Q_1 > 0$ *on* $[\ell - \delta, \ell)$, *then* f *satisfies* (10.45) *if and only if*

$$\int_{\ell-\delta}^{\ell} \frac{a'(s)}{Q_1(s)}\, ds < \infty.$$

(c) *If* $\kappa < 0$, *then* f *satisfies* (10.45) *if and only if*

$$\int_{\ell-\delta}^{\ell} \frac{a'(s)}{\ell - s}\, ds < \infty.$$

This theorem is a corollary of the next one. This in turn follows from Theorems 2.10, 10.1 and 10.26, and, Lemmata 3.5 and 10.20.

Theorem 10.35 (Asymptotic behaviour). *Let* f *be a wavefront solution of* (10.1) *from* ℓ *to* 0 *with wave speed* σ. *Define* Ξ_0 *by* (10.48) *and* Ξ_1 *by* (10.49).

(i) *Suppose that the conditions of Theorem* 10.34 *part* (i) *hold.*

(a) *If* $\kappa > 0$, *then*

$$\frac{(a(f))'(\xi)}{f(\xi)} \to -\frac{\sigma + \sqrt{\sigma^2 - 4(ca')(0)}}{2} \qquad \text{as } \xi \uparrow \Xi_0.$$

(b) *If* $\kappa = 0$, *and,* $Q_0 > 0$ *on* $(0, \delta]$, *then*

$$\frac{(a(f))'(\xi)}{Q_0(f(\xi))} \to -1 \qquad \text{as } \xi \uparrow \Xi_0.$$

(c) *If* $\kappa < 0$, *and,* $c < 0$ *on* $(0, \delta]$, *then*

$$\frac{f'(\xi)}{|c(f(\xi))|} \to -\frac{2}{\sqrt{\sigma^2 - 4(ca)'(0)} - \sigma} \qquad \text{as } \xi \uparrow \Xi_0.$$

(ii) *Suppose that the conditions of Theorem* 10.34 *part* (ii) *hold.*

(a) *If* $\kappa > 0$, *and,* $c > 0$ *on* $[\ell - \delta, \ell)$, *then*

$$\frac{f'(\xi)}{c(f(\xi))} \to -\frac{2}{\sigma + \sqrt{\sigma^2 - 4(ca')'(\ell)}} \qquad \text{as } \xi \downarrow \Xi_1.$$

(b) *If* $\kappa = 0$, *and,* $Q_1 > 0$ *on* $[\ell - \delta, \ell)$, *then*

$$\frac{(a(f))'(\xi)}{Q_1(f(\xi))} \to -1 \qquad \text{as } \xi \downarrow \Xi_1.$$

(c) *If* $\kappa < 0$, *then*

$$\frac{(a(f))'(\xi)}{\ell - f(\xi)} \to -\frac{\sqrt{\sigma^2 - 4(ca')(\ell)} - \sigma}{2} \qquad \text{as } \xi \downarrow \Xi_1.$$

As an illustration of the amalgamated results of this section, let us consider the following fusion of the porous media equation and the Nagumo equation. The lion's share of the conclusions for this example is covered by Application 10.2 and Theorems 10.1, 10.26, and 10.32 – 10.35. The missing details can be deduced from Lemma 10.20 along the lines of the proof of Theorems 10.34 and 10.35.

Application 10.36. *For every $m > 0$ and $0 < \alpha < 1$ the equation*

$$u_t = (u^m)_{xx} + u(1-u)(u-\alpha) \tag{10.66}$$

admits exactly one distinct wavefront solution f from 1 to 0 with a unique wave speed σ and no other wavefront solutions from 1 to 0. The unique wave speed is equal to $(1-2\alpha)/\sqrt{2}$ if $m = 1$, and, for fixed m, is a strictly decreasing and continuous function of α which vanishes for $\alpha = (m+1)/(m+3)$. There holds $f(\xi) < 1$ for all $-\infty < \xi < \infty$, and,

$$\{1 - f(\xi)\}^{-1} f'(\xi) \to -\frac{2m(1-\alpha)}{\sigma + \sqrt{\sigma^2 + 4m(1-\alpha)}} \qquad \textit{as } \xi \to -\infty.$$

(a) *If $m < 1$, then the support of f is not bounded above, and,*

$$f^{(m-3)/2}(\xi) f'(\xi) \to -\sqrt{\frac{2\alpha}{m(m+1)}} \qquad \textit{as } \xi \to \infty.$$

(b) *If $m = 1$, then the support of f is not bounded above, and,*

$$f^{-1}(\xi) f'(\xi) \to -\frac{\sigma + \sqrt{\sigma^2 + 4\alpha}}{2} \qquad \textit{as } \xi \to \infty.$$

(c) *If $m > 1$ and $\alpha < (m+1)/(m+3)$, then the support of f is bounded above, and, defining Ξ_0 by (10.48), there holds*

$$f^{m-2}(\xi) f'(\xi) \to -\frac{\sigma}{m} \qquad \textit{as } \xi \uparrow \Xi_0.$$

(d) *If $m > 1$ and $\alpha = (m+1)/(m+3)$, then the support of f is bounded above, and, defining Ξ_0 by (10.48), there holds*

$$f^{(m-3)/2}(\xi) f'(\xi) \to -\sqrt{\frac{2}{m(m+3)}} \qquad \textit{as } \xi \uparrow \Xi_0.$$

(e) *If $m > 1$ and $\alpha > (m+1)/(m+3)$, then the support of f is not bounded above, and,*

$$f^{-1}(\xi) f'(\xi) \to \frac{\alpha}{\sigma} \qquad \textit{as } \xi \to \infty.$$

A significant area in which reaction-diffusion equations of the form (10.1) arise is combustion theory. In particular the equation

$$u_t = u_{xx} + K(1-u)u^p e^{-E/u},$$

with $K > 0$, p, and, $E > 0$ constants, arises as a model of the flame produced by combustion of a premixed fuel. In this equation, u denotes temperature which has been normalized so that the adiabatic flame temperature is unity, and, the reaction term is given by the Arrhenius law with E the scaled activation energy. Since the term u^p is considered as a weak temperature dependence, the constant p is frequently assumed to be zero. Let $0 < u_0 < 1$ denote the normalized ambient temperature at which the fuel is supplied. Then, per definition, a *plane deflagration flame* is a wavefront solution from 1 to u_0. It transpires however — from the theory of the previous section, for instance — that the above equation cannot admit such a wavefront solution. This phenomenon is often referred to as the cold-boundary difficulty. The modelling device employed to overcome this objection is to postulate an ignition temperature α below which no combustion takes place. This leads [31, 58, 59, 314] to the reaction-diffusion equation

$$u_t = u_{xx} + \begin{cases} 0 & \text{for } 0 \leq u < \alpha \\ K(1-u)u^p e^{-E/u} & \text{for } \alpha \leq u \leq 1. \end{cases} \tag{10.67}$$

Application 10.37. *For any $K > 0$, p, $E > 0$, and, $0 \leq u_0 < \alpha < 1$, there exists a unique wave speed σ such that equation* (10.67) *admits a wavefront solution from 1 to u_0. Moreover, for this wave speed, there is precisely one distinct wavefront solution f from 1 to u_0. The unique wave speed is a positive continuous function of K, p, E, u_0 and α, and, if the remaining variables are fixed, is a strictly increasing function of K and u_0, and, a strictly decreasing function of p and E, respectively. The wavefront f is such that $1 > f(\xi) > u_0$ for all $-\infty < \xi < \infty$,*

$$\{1 - f(\xi)\}^{-1} f'(\xi) \to -\frac{2Ke^{-E}}{\sigma + \sqrt{\sigma^2 + 4Ke^{-E}}} \quad \textit{as } \xi \to -\infty$$

and

$$\{f(\xi) - u_0\}^{-1} f'(\xi) \to -\sigma \quad \textit{as } \xi \to \infty.$$

Note that the conclusions of this example apply without change to the commonly-used [32, 43] 'linearization' of equation (10.67)

$$u_t = u_{xx} + \begin{cases} 0 & \text{for } 0 \leq u < \alpha \\ Ke^{-E}(1-u)u^p e^{-E(1-u)} & \text{for } \alpha \leq u \leq 1. \end{cases}$$

The fact that these models admit a wavefront solution with a unique positive wave speed is of particular importance in the description of combustion processes. It means that by supplying the combustive fuel mixture at the appropriate rate in a counterflow, a steady flame can be sustained.

10.3 Smooth coefficients

Having dealt with reaction-diffusion equations of the form (10.1), where the reaction term has a fixed sign or at worst one change of sign, but where nothing beyond the basic assumptions in Hypothesis 2.1 has been assumed about the smoothness of the coefficients, in this section we shall discuss the more general situation in which the coefficients have a prescribed smoothness. This smoothness is exhibited by the prototypes of the Fisher equation, the Newell–Whitehead equation, the Zeldovich equation, the KPP equation, and, the Nagumo equation.

The following is the main result of this section.

Theorem 10.38. *Suppose that $\ell < \infty$ and ca' is differentiable in $(0, \ell)$. Let S denote the set of wave speeds for which equation (10.1) has a wavefront solution from ℓ to 0. Then equation (10.1) has exactly one distinct wavefront solution from ℓ to 0 with every wave speed $\sigma \in S$. Furthermore, if $ca' \in C^1(I)$ and $(ca')(0) = (ca')(\ell) = 0$, the following alternatives are mutually exclusive.*

(a) *If $c(u) > 0$ for all $0 < u < \ell$, then $S = [\sigma^*, \infty)$ for some $\sigma^* > 0$.*

(b) *If $c(u) < 0$ for all $0 < u < \ell$, then $S = (-\infty, \sigma^*]$ for some $\sigma^* < 0$.*

(c) *If $c(u) > 0$ for all $0 < u < \alpha$, $c(\alpha) = c(\beta) = 0$, and, $c(u) > 0$ for all $\beta < u < \ell$, for some $0 < \alpha \le \beta < \ell$, then either S is empty or $S = [\sigma^*, \sigma^{**})$ for some $0 < \sigma^* < \sigma^{**}$.*

(d) *If $c(u) < 0$ for all $0 < u < \alpha$, $c(\alpha) = c(\beta) = 0$, and, $c(u) < 0$ for all $\beta < u < \ell$, for some $0 < \alpha \le \beta < \ell$, then either S is empty or $S = (\sigma^{**}, \sigma^*]$ for some $\sigma^{**} < \sigma^* < 0$.*

(e) *If $c(u_i^-) \le 0$ for a sequence of values $\{u_i^-\}_{i=1}^\infty \subset (0, \ell)$ such that $u_i^- \to 0$ as $i \to \infty$, and, $c(u_i^+) \ge 0$ for a sequence of values $\{u_i^+\}_{i=1}^\infty \subset (0, \ell)$ such that $u_i^+ \to \ell$ as $i \to \infty$, then either S is empty or $S = \{\sigma^*\}$ for a single value σ^*.*

(f) *If $c(u) > 0$ for all $0 < u < \delta$, and, $c(u_i^+) \le 0$ for a sequence of values $\{u_i^+\}_{i=1}^\infty \subset (0, \ell)$ such that $u_i^+ \to \ell$ as $i \to \infty$, or, if $c(u_i^-) \ge 0$ for a sequence of values $\{u_i^-\}_{i=1}^\infty \subset (0, \ell)$ such that $u_i^- \to 0$ as $i \to \infty$, and, $c(u) < 0$ for all $\ell - \delta < u < \ell$, for some $0 < \delta < \ell$, then S is empty.*

It may be of interest to note that, according to part (c) of Theorem 10.38, if the hypothesis $c(u) > 0$ for all $0 < u < \ell$ is merely violated at a single point $u = \alpha$ with $0 < \alpha < \ell$, this is enough to destroy the conclusion of part (a) of the theorem. A similar remark applies to parts (b) and (d). By implication this also affects the existence results in Section 10.1. In an explicit example, where $c(u) > 0$ for all $0 < u < \alpha$, $c(\alpha) = 0$, and, $c(u) > 0$ for all $\alpha < u < \ell$, for some $0 < \alpha < \ell$, we shall see below that both alternatives in parts (c) are viable. Using a more complicated explicit example we shall also show that both alternatives in parts (d) and (e) can occur. In general, for cases (c) – (e), necessary conditions for the existence of a wavefront solution are provided by Theorems 3.4, 8.12 and 10.1.

Sufficient conditions are given by Theorems 3.4 and 8.13. For case (e) further specific conditions can be found in Theorems 10.1 and 10.26.

Proof of Theorem 10.38. Suppose to begin with that for some $\sigma \in S$, equation (10.1) has two distinct wavefront solutions from ℓ to 0. Then by Theorem 2.10 and Lemmata 2.26 and 2.40, equation (10.3) has two different solutions θ_1 and θ_2 on $[0,\ell]$ which are positive on $(0,\ell)$ and satisfy $\theta_1(\ell) = \theta_2(\ell) = 0$. Moreover, by Lemma 4.7, either $\theta_1 \geq \theta_2$ on $[0,\ell]$ or $\theta_2 \geq \theta_1$ on $[0,\ell]$. However, by (10.7) this implies that $\theta_1 \equiv \theta_2$ or $\sigma = 0$. Subsequently, (10.5) implies that $\theta_1 \equiv \theta_2$ whatever the value of σ. Thus, we have a contradiction. This proves the first assertion of the theorem. To prove the remainder of the theorem, we consider the cases (a), (c), (e) and (f) in turn. The cases (b) and (d) follow from (a) and (c) respectively via Theorem 3.4. We leave it to the reader to check that (a) – (f) are mutually exclusive.

(a) This is just a restatement of Corollary 10.10.

(c) By Corollary 10.10 there exists a $\sigma_0 > 0$ such that equation (10.1) has a wavefront solution from α to 0 for every wave speed $\sigma \geq \sigma_0$ and no such solution for any wave speed $\sigma < \sigma_0$. Similarly, by a simple redefinition of the dependent variable, $\widetilde{u} := u - \beta$ say, there exists a $\sigma_1 > 0$ such that (10.1) has a wavefront solution from ℓ to β for every wave speed $\sigma \geq \sigma_1$ and no such solution for any wave speed $\sigma < \sigma_1$. It follows from Theorem 8.12 part (i) that $S \subseteq [\sigma_0, \infty)$, and, from Theorem 8.12 part (ii) that $S \subseteq (-\infty, \sigma_1)$. Theorem 8.3(iv) then gives the result.

(e) Suppose that S contains two values σ_1 and $\sigma_2 \geq \sigma_1$. Then by Theorem 2.10 and Lemma 2.40 there exists a solution θ_i of (10.3) with $\sigma = \sigma_i$ on $[0,\ell]$ which is positive on $(0,\ell)$ and such that $\theta_i(\ell) = 0$ for $i = 1, 2$. Simultaneously, the function Θ_i, defined by $\Theta_i(s) := \theta_i(\ell - s)$ for $0 \leq s \leq \ell$, is a solution of (10.16) with $\sigma = \sigma_i$ on $[0,\ell]$ with similar properties. However, by Lemmata 2.14(i), 2.18, 4.7 and 6.12, there holds $\theta_2 \geq \theta_1$ on $[0,\ell)$; while, by the same token, $\Theta_2 \geq \Theta_1$ on $[0,\ell)$. This is clearly incompatible unless $\theta_1 \equiv \theta_2$ and $\sigma_1 = \sigma_2$.

(f) If $c > 0$ on $(0,\delta)$ for some $0 < \delta < \ell$, Theorem 10.1 says that every $\sigma \in S$ is necessarily positive. On the other hand, if $c(u_i^+) \leq 0$ for a sequence of values $\{u_i^+\}_{i=1}^{\infty} \subset (0,\ell)$ such that $u_i^+ \to \ell$ as $i \to \infty$, by Lemmata 3.5 and 6.12, equation (10.16) has a solution only if $\sigma \leq 0$. Subsequently, invoking Theorem 2.10 and Lemma 3.5, if both hypotheses hold, S must be empty. Similarly, if $c < 0$ on $(\ell - \delta, \ell)$, for some $0 < \delta < \ell$, every $\sigma \in S$ has to be negative; while, if $c(u_i^-) \geq 0$ for a sequence of values $\{u_i^-\}_{i=1}^{\infty} \subset (0,\ell)$ such that $u_i^- \to 0$ as $i \to \infty$, equation (10.3) has a solution only if $\sigma \geq 0$. So S is empty in this case too. □

As illustrations of Theorem 10.38 let us consider the following two examples. These aptly illustrate the strength of the integral equation approach for determining the existence of travelling-wave solutions as compared to the more time-honoured method of phase-plane analysis.

Application 10.39. *Consider the equation*

$$u_t = u_{xx} + \begin{cases} u(\alpha - u)^2 & \text{for } 0 \leq u \leq \alpha \\ k(\ell - u)(u - \alpha)^2 & \text{for } \alpha < u \leq \ell, \end{cases} \tag{10.68}$$

where $0 < \alpha < \ell$ *and* k *are real parameters.*

(a) *If* $k \leq 8\alpha^2/(\ell - \alpha)^2$, *then equation* (10.68) *admits no wavefront solutions from* ℓ *to* 0.

(b) *If* $k > 8\alpha^2/(\ell - \alpha)^2$, *then there exists a* $2\alpha \leq \sigma^* < \sqrt{k/2}(\ell - \alpha)$ *such that* (10.68) *admits exactly one distinct wavefront solution from* ℓ *to* 0 *with wave speed* σ *for every* $\sigma^* \leq \sigma < \sqrt{k/2}(\ell - \alpha)$ *and no such solution for any other wave speed.*

Proof. In the light of Theorem 10.38(f), it suffices to restrict attention to $k > 0$. In this case, by Corollary 10.10, equation (10.68) has a wavefront solution from α to 0 with wave speed σ if and only if $\sigma \geq 2\alpha$; while, in terms of the dependent variable $\widetilde{u} := u - \alpha$, Application 10.14 shows that the equation has a wavefront solution from ℓ to α with wave speed σ if and only if $\sigma \geq \sigma_1 := \sqrt{k/2}(\ell - \alpha)$. Theorem 8.12 part (i) subsequently implies that if (10.68) has a wavefront solution from ℓ to 0 with wave speed σ, then necessarily $\sigma \geq 2\alpha$. Whereas, recalling that by Theorem 10.38, wavefront solutions of (10.68) are distinct, Theorem 8.12 part (ii) implies that $\sigma < \sigma_1$. Alternatively, by Theorem 8.13, given any $2\alpha \leq \sigma_0 < \sigma_1$ there exists a $\sigma_0 \leq \sigma \leq \sigma_1$ such that (10.68) has a wavefront solution from ℓ to 0 with wave speed σ. Combining these deductions with Theorem 10.38(c) provides the desired conclusions. □

Application 10.40. *Consider the equation*

$$u_t = u_{xx} + \begin{cases} u(u - \alpha)(\beta - u)^2 & \text{for } 0 \leq u \leq \beta \\ k(\ell - u)(u - \beta)^2 & \text{for } \beta < u \leq \ell, \end{cases} \tag{10.69}$$

where $0 < \alpha < \beta < \ell$ *and* k *are real parameters.*

(a) *If* $5\alpha < 2\beta$, *then there exists a* $k^* > 0$, *which depends only on* α, β *and* ℓ, *with the following property. When* $k \leq k^*$ *equation* (10.69) *admits no wavefront solutions from* ℓ *to* 0. *When* $k > k^*$, *there exists a number* $\sigma^* > 0$ *such that* (10.69) *admits exactly one distinct wavefront solution from* ℓ *to* 0 *with wave speed* σ^* *and no such solution with any other wave speed.*

(b) *If* $5\alpha = 2\beta$, *then equation* (10.69) *admits no wavefront solutions from* ℓ *to* 0 *when* $k \leq 0$. *When* $k > 0$, *there exists a number* $\sigma^* > 0$ *such that* (10.69) *admits exactly one distinct wavefront solution from* ℓ *to* 0 *with wave speed* σ^* *and no such solution with any other wave speed.*

(c) *If* $5\alpha > 2\beta$, *then there exists a* $k^* < 0$, *which depends only on* α, β *and* ℓ, *with the following property. When* $k \leq k^*$, *equation* (10.69) *admits no wavefront solutions from* ℓ *to* 0. *When* $k^* < k < 0$, *there exist numbers* $\sigma^{**} < \sigma^* < 0$ *such that* (10.69) *admits exactly one distinct wavefront solution from* ℓ *to* 0 *with any wave speed* $\sigma^{**} < \sigma \leq \sigma^*$ *and no such solution with any other wave speed. When* $k \geq 0$, *there exists a single number* σ^* *such that* (10.69) *admits exactly one distinct wavefront solution from* ℓ *to* 0 *with wave speed* σ^* *and no such solution with any other wave speed. In this event,* $\sigma^* < 0$ *if* $k < \beta^3(5\alpha - 2\beta)/5(\ell - \beta)^4$, $\sigma^* = 0$ *if* $k = \beta^3(5\alpha - 2\beta)/5(\ell - \beta)^4$, *and,* $\sigma^* > 0$ *if* $k > \beta^3(5\alpha - 2\beta)/5(\ell - \beta)^4$.

Proof. Let S denote the set of wave speeds for which equation (10.69) has a wavefront solution from ℓ to 0. By Corollary 10.27, there exists a number σ_0, which depends only on α and β, such that (10.69) has a wavefront solution from β to 0 with wave speed σ_0 and no other wave speed. Moreover, by Theorem 10.1, $\sigma_0 > 0$ if $5\alpha < 2\beta$, $\sigma_0 = 0$ if $5\alpha = 2\beta$, and, $\sigma_0 < 0$ if $5\alpha > 2\beta$. Theorem 8.12 part (i) implies that $S \subseteq (\sigma_0, \infty)$. To proceed, we distinguish according to the sign of k. However, first we observe that by Theorem 2.10, Lemmata 2.14(i), 2.18, 2.40 and 4.7, and, the aforesaid; equation (10.1) has a unique solution $\theta(\cdot;\sigma)$ on $[0,\beta]$ which is positive on $(0,\beta)$ for every $\sigma \geq \sigma_0$. Moreover, $\theta(\beta;\sigma_0) = 0$, and, by Lemmata 4.7 and 8.5, $\sigma \mapsto \theta(\beta;\sigma)$ is a strictly increasing, continuous function on $[\sigma_0, \infty)$.

(i) Suppose that $k > 0$. Then, by the argument in the previous example, equation (10.69) has a wavefront solution from ℓ to β with every wave speed $\sigma \geq \sigma_1 := \sqrt{k/2}(\ell - \beta)$ and no other wave speed. Hence, by Theorem 8.12 part (ii), $S \subseteq (-\infty, \sigma_1)$. This implies that S is empty if $\sigma_1 \leq \sigma_0$. On the other hand, if $\sigma_1 > \sigma_0$, by Theorem 8.13 there exists a $\sigma_0 \leq \sigma^* \leq \sigma_1$ such that $\sigma^* \in S$. Recalling Theorems 10.1 and 10.38(e) this gives the desired results.

(ii) Suppose that $k = 0$. In this case, we note that any solution of equation (10.16) must have the form $\Theta(s;\sigma) := -\sigma s$ on $[0, \ell - \beta]$. Subsequently, by Theorem 2.10 and Lemma 8.9, $\sigma \in S$ if and only if $\sigma_0 < \sigma < 0$ and $\theta(\beta;\sigma) = \Theta(\ell - \beta;\sigma) > 0$. This is plainly not possible when $\sigma_0 \geq 0$. On the other hand when $\sigma_0 < 0$, elementary analysis shows that there is a unique value σ^* for which this holds.

(iii) Suppose that $k < 0$. Then, by assigning $\widetilde{u} := \ell - u$ to be the dependent variable, Corollary 10.10 tells us that (10.68) has a wavefront solution from ℓ to β with wave speed σ if and only if $\sigma \leq \sigma_1 := -2\sqrt{-k}(\ell - \beta)$. Hence, by Theorem 8.12(ii), $S \subseteq (-\infty, \sigma_1]$. It follows that if $\sigma_1 \leq \sigma_0$, then S is empty. Suppose therefore that $\sigma_1 > \sigma_0$. Then, by Lemmata 2.17 and 2.18, equation (10.16), has a unique maximal solution $\Theta(\cdot;\sigma)$ on $[0, \ell - \beta]$ which is positive on $(0, \ell - \beta]$ for every $\sigma \leq \sigma_1$. Moreover, by Lemmata 2.18 and 8.5, $\sigma \mapsto \Theta(\ell - \beta;\sigma)$ is a decreasing function which is continuous from the left on $(-\infty, \sigma_1]$. It follows that we can define a $\sigma^* > \sigma_0$ by

$\sigma^* := \sup\{\sigma \in (\sigma_0, \sigma_1] : \theta(\beta;\sigma) \leq \Theta(\ell - \beta;\sigma)\}$. In which case, by Lemma 4.8, for any $\sigma_0 < \sigma \leq \sigma^*$ there exists a unique solution Θ of equation (10.16) on $[0, \ell - \beta]$ such that $\Theta(\ell - \beta) = \theta(\beta;\sigma) > 0$. Theorem 2.10 and Lemma 8.9 then yield $S = (\sigma_0, \sigma^*]$. □

With regard to the set S of wave speeds for which an equation of the type (10.1) satisfying the hypotheses Theorem 10.38 may admit a wavefront solution from ℓ to 0, Application 10.39 shows that both alternatives in Theorem 10.38 part (c) can occur dependent on the size of the parameter $k > 0$. Correspondingly, Application 10.40 with $k < 0$ shows that both alternatives in part (d) are possible, and, with $k > 0$ that both alternatives in Theorem 10.38 part (e) may occur. Incidentally, Application 10.40 with $k > 0$ also shows that if, in Theorem 10.26(a) and Corollary 10.27, the condition $c(u) > 0$ for all $\alpha < u < \ell$ is violated at just a single point $u = \beta$, $\alpha < \beta < \ell$, the stated result need no longer be true.

Bibliographic notes

The importance of wavefront solutions of reaction-diffusion equations of the class (10.1) can be gauged from [25,56,82,101,134,170,210,211,296,297,306,315]. Among the many results obtained specific reference can be made to those of Aronson and Weinberger [15, 20, 21], Atkinson et al. [22], Berestycki et al. [42, 43], Fife and McLeod [102–104], Grindrod and Sleeman [135], Hadeler [141–145], Hadeler and Rothe [150], Hosono [163], McKean [201], de Pablo and Sánchez [226], de Pablo and Vázquez [228], Pauwelussen and Peletier [232], Sánchez-Garduño and Maini [254,256,257], Uchiyama [284,286], and, Vol'pert et al. [297]. Many of these results have also been extended to a higher number of dimensions [29, 44, 45, 108].

Theorem 10.1 is well-known for the semilinear version of equation (10.1) [21, 82, 100, 101, 103, 143, 211, 296, 297]. The main conclusions of this theorem are classically obtained directly from the ordinary differential equation $(a(f))'' + c(f) + \sigma f' = 0$ for a travelling-wave solution $u = f(x - \sigma t)$ of (10.1) [82, 143, 211]. That all the conclusions can be easily deduced from the integral equation, follows from ideas in [21, 101, 103, 297].

In the case that $p = 1$, equation (10.8) is the Nagumo equation in one of its many guises, and for this equation Application 10.2 is far from new. The critical wave speed in this case was obtained earlier by Hadeler and Rothe [150]. See also [82,141,142,170,211]. Moreover, according to McKean [201] and Sattinger [263] the explicit solution (10.9) with $p = 1$ was found by Huxley.

For the traditional estimate (10.26) of the critical wave speed σ^* in Theorem 10.5(b), improved in Theorem 10.8, see [20, 21, 56, 82, 101, 142, 143, 150, 203, 284, 296, 297]. Furthermore, the well-known results on the existence of wavefront solutions of the KPP equation can be found in [78, 186], and, a direct extension to reaction-convection-diffusion equations in [197]. The proof of the inequality $\sigma^* < \sqrt{2\Lambda_2}$ included in Theorem 10.8 has its genesis in [22].

The distinction between a pushed and pulled wave originally made by Stokes [276], was also used in [94, 232, 286, 291]. Information on the marginal stability principle is contained in [40, 79, 241, 250–252].

In the case $p = 1$, the Application 10.14 has been considered in [82, 141, 142, 150, 192, 211, 268]. The consequences of this example for the Fisher, Newell–Whitehead and Zeldovich equations results are well-known [56,78,82,101,105,134, 186, 210, 211]. In a number of specific cases of Application 10.14 the critical wave speed has been identified using the marginal stability principle [40, 79, 251, 252]. In chronological order, this has been achieved when $p = 1$ and $c_0 = c_1$ by Dee and Langer [79], when $p = 1$ and $c_0c_1 > 0$ by Ben-Jacob et al. [40], and, when $p = 2$ and $c_1 = c_0\ell^4$ by van Saarloos [252]. The case $p = 2$ and $c_1 = 1$ was also considered in [242], and, the case that $2p$ is a positive integer and $c_1 = 1$ in [240].

Theorem 10.16 generalizes results of Pauwelussen and Peletier [232] and of Vol'pert et al. [297]. Theorem 10.17 improves on results of Hadeler [143–145] and of Pauwelussen and Peletier [232]. Theorem 10.18 expands on the work of Benguria and Depassier [33, 35–38] and of Benguria et al. [32]. Indeed, the most essential ideas behind the proof of part (a) can be found in [32, 33, 35, 36, 38] and part (b) in [37]. Part (a) has been extended to equations of the form $u_t = u_{xx} + b_0(u_x)^2/(1-u) + c(u)$, where b_0 is a constant, in [39]

Significant contributions to the question of whether or not wavefront solutions of degenerate nonlinear reaction-diffusion equations exhibit finite speed of propagation, which is succinctly covered in Theorem 10.21, have been made in chronological order by Aronson [15], Atkinson et al. [22], Grindrod and Sleeman [135], de Pablo and Vázquez [228], Sánchez-Garduño and Maini [254,256,257], and, de Pablo and Sánchez [226]. The difficulty that the degeneration of the equation causes for the numerical computation of travelling waves is touched upon in [206].

Prior studies to Theorem 10.22 of the asymptotic behaviour $f(\xi)$ as $\xi \to \pm\infty$ of a wavefront solution f to the semilinear version of equation (10.1) were conducted by Uchiyama [284], by Vol'pert [298], and by Vol'pert et al. [297]. It can be verified that Theorem 10.22 covers, and in some instances improves on, their results. Some additional analysis of the higher order asymptotic behaviour of a wavefront solution can be found in [22, 284, 296, 298]. A particular case is also examined in [242].

Theorem 10.23 builds on the work of Kelley [180].

Application 10.24 has been investigated in the past for $m \geq 1$ and $p = q = 1$ by Aronson [15] and by Atkinson et al. [22], for $m = 1$, $p = 2$ and $q = 1$ by Billingham and Needham [47], for $m > 1$ and $q = 1$ by de Pablo and Vázquez [228], for $m > 1$, $p < 1$ and $q = 1$ by de Pablo and Sánchez [226] and by Medvedev et al. [203], for $m > 1$ and $p \geq 1$ by Biró [48, 49], and, for $m = 1$, $p \geq 1$ and $q = 1$ by Needham and Barnes [215]. In particular, for the case that $q = 1$ is fixed but $m > 1$ and $p < 1$ may vary, de Pablo and Sánchez [226] have shown that the critical wave speed σ^* is an analytic function of $m + p \geq 2$. Their proof is based on the theory of anomalous exponents developed by Aronson and Vázquez [18, 19], who discussed the case with $p = q = 1$ and variable m in [18,19]. Application 10.25 in the case $m = 1 \geq n \geq 0$ was also studied by Atkinson et al. [22].

As mentioned in the main text, Corollary 10.31 was essentially deduced by Zel'dovich [314]. It has been proven under more restrictive regularity assumptions on the functions a and c than those imposed in this monograph by Berestycki et al. [42,43]. The proof of Theorem 10.30 has essentially been obtained by amalgamating ideas in [42,43] with the study of the integral equation (10.3).

Earlier results of the same ilk as Theorem 10.32 were obtained by Pauwelussen and Peletier [232] and by Vol'pert et al. [297]. Theorem 10.32 includes these results. Previous results similar to Theorem 10.33 were obtained by Hadeler [143–145]. These results can be seen to fall under Theorem 10.33 if one invokes Lemma 10.20 to identify the behaviour of the function $\theta^*(u)$ as $u \downarrow 0$ and $u \uparrow \ell$.

Theorem 10.34 summarizes the results of Hosono [163], of Grindrod and Sleeman [135], of Wang [300], and, of Sánchez-Garduño and Maini [257], for which Application 10.28 formed the prototype equation, in a general setting. Previous authors who have investigated the asymptotic behaviour of wavefront solutions of reaction-diffusion equations where the reaction term has one change of sign are Aronson and Weinberger [21], Berestycki et al. [42, 43], Uchiyama [286], Hosono [163], Vol'pert [298], and, Vol'pert et al. [297].

In [135] it is stated that equation (10.66) with $m = 1$ and $0 < \alpha < 1$ admits a wavefront solution from 1 to 0 if and only if $\alpha \leq 1/2$. Application 10.36 rectifies this statement. The condition $\alpha \leq 1/2$ is necessary and sufficient for the equation to admit a wavefront solution from 1 to 0 with a nonnegative wave speed. The confusion presumably arises because the main interest in [135] was in establishing necessary and sufficient conditions for (10.66) to admit a wavefront solution displaying finite speed of propagation in the case $m = 2$. Application 10.36 shows that equation (10.66) with $m > 0$ and $0 < \alpha < 1$ admits a wavefront solution f from 1 to 0 such that the support of f is bounded above, if, and only if, $m > 1$ and $\alpha \leq (m+1)/(m+3)$. In turn, the condition $\alpha \leq (m+1)/(m+3)$ is equivalent to the conclusion that the unique wave speed σ is nonnegative.

For more information on the mathematics of combustion models with ignition thresholds, we refer to [42, 43]. In particular, in [42, 43], Berestycki et al. have investigated the high activation energy limit of wavefront solutions of a more general class of combustion models than in Application 10.37, viz. $u_t = (a(u))_{xx} + c(u; E)$, where $a \in C^1([0,1])$, $a'(u) > 0$ for all $0 \leq u \leq 1$, and $c(u; E) = 0$ for $0 \leq u < \alpha < 1$

To complete the bibliographical notes on reaction-diffusion equations with one sign change in the reaction term, we refer to [209], where de Mottoni studied travelling-wave solutions of an equation of the form (10.1) in which the reaction term has a single sign change and the diffusion term $a'(u)$ vanishes for a connected range of values of u.

Application 10.28 is due to Hadeler and Rothe [150]. This result can also be found in [82,141,142]. This particular example ably illustrates the possibility of a single reaction-diffusion equation admitting wavefronts connecting different zeros of the reaction term. In [296–298] a family of such solutions is called a *system of waves*. The possibility that a more general equation of the form $u_t = u_{xx} + c(u)$ with $c \in C^1([0,\ell])$ and $c(0) = c(\alpha) = c(\ell) = 0$ for some $0 < \alpha < \ell < \infty$ admit wavefront solutions from α to 0, from ℓ to α, and, from ℓ to 0 has been studied in some detail in [102–104, 297]. In particular, among other results, Fife and McLeod [102] proved that if $c \leq 0$ in a right neighbourhood of zero, then any wavefront solution from ℓ to 0 necessarily moves faster than one from α to 0. Likewise, if $c \geq 0$ in a left neighbourhood of ℓ, then any wavefront solution from ℓ to α necessarily moves faster than one from ℓ to 0. These conclusions are borne out by the example of the Nagumo equation, where $\sigma_0^* < \sigma^* < \sigma_1^*$ for every relevant value of α, and, have been generalized in Theorem 8.12.

In [103] one may find a converse to Theorem 8.12. This says that should the equation $u_t = u_{xx} + c(u)$ admit a wavefront solution from α to 0 with wave speed σ_0 and one from ℓ to α with wave speed $\sigma_1 > \sigma_0$, then it would admit a wavefront solution from ℓ to 0 with some wave speed σ, where $\sigma_0 < \sigma < \sigma_1$. With a heuristic argument as basis, this statement can also be found in [270]. In addition, under the supplementary hypotheses that $c(u) \leq 0$ for all $0 < u < \delta$ and $c(u) \geq 0$ for all $\ell - \delta < u < \ell$, for some $0 < \delta < \ell/2$, it is stated as a theorem in [296]. Unequivocally though, Application 10.40 provides a counterexample to this proposition. When $k > 0$, equation (10.69) is such that $c(u) < 0$ for all $0 < u < \alpha$ and $c(u) > 0$ for all $\beta < u < \ell$. Moreover, if $5\alpha < 2\beta$ there exists a unique wave speed $\sigma_0 > 0$ such that the equation has a wavefront solution from β to 0, and, there exists a $\sigma_1^* > 0$ such that (10.69) has a wavefront solution from ℓ to β with wave speed σ if and only if $\sigma \geq \sigma_1^*$. Thus for every $k > 0$ it is possible to choose a $\sigma_1 > \sigma_0$ such that equation (10.69) has a wavefront solution from β to 0 with wave speed σ_0 and one from ℓ to β with wave speed σ_1. However, if k is sufficiently small, the equation does not admit any wavefront solution from ℓ to 0 whatsoever.

The assertion to which equation (10.69) provides a counterexample is justified in [103] with an argument based on our integral equation in differential form. The basis of the argument is that if $u_t = u_{xx} + c(u)$ has a wavefront solution from α to 0 with wave speed σ_0 and one from ℓ to α with wave speed $\sigma_1 > \sigma_0$, then for every $\sigma \geq \sigma_0$ there is a solution θ of the equation (10.3) on $[0, \alpha]$ and for every $\sigma \leq \sigma_1$ there is a solution Θ of (10.16) on $[0, \ell - \alpha]$. Furthermore, since $\theta(\alpha) = 0$ for a solution of (10.3) with $\sigma = \sigma_0$ and the maximal solution of (10.3) is a nondecreasing function of σ, while $\Theta(\ell - \alpha) = 0$ for a solution of (10.16) with $\sigma = \sigma_1$ and the maximal solution of (10.16) is a nonincreasing function of σ, by continuity there must be a $\sigma_0 < \sigma < \sigma_1$ for which a match $\theta(\alpha) = \Theta(\ell - \alpha)$ occurs. In this event, Lemma 8.9 would yield the desired conclusion. The aspect that is overlooked is that this match may occur exclusively if $\theta(\alpha) = \Theta(\ell - \alpha) = 0$.

The converse of Theorem 8.12 is true under a suitable modification of the more general statement in [103]. See Theorem 8.13 for details.

Theorem 10.38 extends, sharpens, and clarifies the theory of wavefront solutions of scalar semilinear reaction-diffusion equations as developed by Vol'pert et al. in [296, 297]. A possible tool for shedding light on which of the alternatives in parts (c) – (e) may apply is the method of directed graphs expounded by Terman [281, 282].

Chapter 11

Unbounded waves

One of the motivations for studying travelling-wave solutions of equations of the class (1.1) is that these solutions may be used as a tool for determining the properties of an arbitrary solution of the partial differential equation. As one illustration we mention the characterization of finite speed of propagation using semi-wavefront solutions in [126, 127]; as another, the characterization of complete and incomplete blow-up in diffusion-reaction processes dependent upon whether or not the equation admits an unbounded strict semi-wavefront solution in [114].

In this chapter we shall summarize some elementary properties of unbounded travelling-wave solutions of equations of the type (1.1) which can be derived simply from study of the integral equation (1.9). We recall that an unbounded monotonic travelling-wave solution of equation (1.1) decreasing to 0 is a solution of the form $u = f(x - \sigma t)$ where σ is a real number, and, f is a function which is defined and monotonic in some domain $\Omega = (\omega, \infty)$ with $-\infty \leq \omega < \infty$, $f(\xi) \to 0$ as $\xi \to \infty$, and, $f(\xi) \to \infty$ as $\xi \downarrow \omega$. If $\omega > -\infty$, we say that f is a strict semi-wavefront solution. If $\omega = -\infty$, it is global. In both cases, σ is the wave speed.

Definition 11.1. Suppose that $\ell = \infty$. The partial differential equation (1.1) will be said to admit a one-parameter family of distinct unbounded monotonic travelling-wave solutions with wave speed σ decreasing to 0 when there exists a continuous order-preserving bijective mapping from the interval $[0,1]$ onto the set of all such solutions.

Theorem 11.2. *Suppose that $\ell = \infty$ and that $c(u) < 0$ for all $u > 0$. Then equation (1.1) has exactly one distinct unbounded monotonic travelling-wave solution decreasing to 0 for every wave speed σ.*

Theorem 11.3. *Suppose that $\ell = \infty$ and $c(u) > 0$ for all $u > 0$. Then the set of wave speeds for which equation (1.1) has an unbounded monotonic travelling-wave solution decreasing to 0 is either empty or an interval of the form $[\sigma^*, \infty)$ for finite σ^*. Moreover, in the latter case (1.1) has a one-parameter family of distinct unbounded monotonic travelling-wave solutions decreasing to 0 in the sense of Definition 11.1 with every wave speed $\sigma > \sigma^*$, and, exactly one such solution or a one-parameter family of these solutions with wave speed σ^*.*

Theorem 11.4. *Consider equation (1.1) with two sets of coefficients a_i, b_i and c_i on $[0, \infty)$ for $i = 1, 2$. Let σ_1 and σ_2 denote real parameters.*

(a) *Suppose that $u \mapsto \sigma_2 u + b_2(u) - \sigma_1 u - b_1(u)$ is a nondecreasing function on $(0, \infty)$, and, $(c_2 a_2')(u) \leq (c_1 a_1')(u)$ for all $u > 0$.*

(b) *Suppose that* $\sigma_2 u + b_2(u) \geq \sigma_1 u + b_1(u)$ *and* $\max\{0, (c_2 a_2')(u)\} \leq (c_1 a_1')(u)$ *for all* $u > 0$.

(c) *Suppose that in addition to the hypotheses of (b) there holds* $\sigma_2 u + b_2(u) > \sigma_1 u + b_1(u)$ *for some* $u > 0$ *and* $(c_2 a_2')(u) > 0$ *for all* $u > 0$.

Then in both cases (a) *and* (b), *if equation* (1.1) *with* $i = 1$ *admits an unbounded monotonic travelling-wave solution with speed* σ_1 *decreasing to* 0, *so does* (1.1) *with* $i = 2$ *and speed* σ_2. *In particular, in case* (c), *the last-mentioned equation admits a one-parameter family of such solutions. Idem ditto, if equation* (1.1) *with* $i = 1$ *admits an unbounded strict semi-wavefront solution with speed* σ_1 *decreasing to* 0, *so does* (1.1) *with* $i = 2$ *and speed* σ_2. *Furthermore, in case* (c), *the last-mentioned equation admits a one-parameter family of such solutions.*

Theorem 11.5. *Suppose that* $\ell = \infty$. *Then the equation*

$$u_t = (a(u))_{xx} + (b(u))_x \tag{11.1}$$

admits an unbounded monotonic travelling-wave solution decreasing to 0 *if and only if*

$$\sup_{0<u<\infty} \left\{ -\frac{b(u)}{u} \right\} < \infty. \tag{11.2}$$

Furthermore, the equation admits an unbounded strict semi-wavefront solution decreasing to 0 *if and only if* (11.2) *holds and*

$$\int_1^\infty \frac{a'(s)}{\max\{s, b(s)\}} \, ds < \infty.$$

Theorem 11.6. *Suppose that* $\ell = \infty$ *and* $c(u) \leq 0$ *for all* $u > 0$. *Then the equation*

$$u_t = (a(u))_{xx} + c(u) \tag{11.3}$$

admits an unbounded monotonic travelling-wave solution decreasing to 0. *Furthermore, the equation admits an unbounded strict semi-wavefront solution decreasing to* 0 *if and only if*

$$\int_1^\infty \frac{a'(s)}{\max\{s, Q(s)\}} \, ds < \infty, \tag{11.4}$$

where

$$Q(s) = \left| 2 \int_0^s c(r) a'(r) \, dr \right|^{1/2}. \tag{11.5}$$

Theorem 11.7. *Suppose that* $\ell = \infty$ *and* $c(u) \geq 0$ *for all* $u > 0$. *Then equation* (11.3) *admits an unbounded monotonic travelling-wave solution decreasing to* 0 *if and only if*

$$\sup_{0<s<\infty} \left\{ \frac{1}{s} \int_0^s \frac{c(u) a'(u)}{u} \, du \right\} < \infty. \tag{11.6}$$

Furthermore, the equation admits an unbounded strict semi-wavefront solution decreasing to 0 *if and only if* (11.6) *holds and*

$$\int_1^\infty \frac{a'(s)}{s}\,ds < \infty. \tag{11.7}$$

At the risk of stating the obvious, if the conclusion of any of the last three theorems is that the equation admits an unbounded monotonic travelling-wave solution decreasing to 0, but, it does not admit an unbounded strict semi-wavefront solution decreasing to 0, then *every* solution in the first-mentioned category is necessarily global.

In the next chapter we shall complement the above results with a complete characterization of global monotonic travelling-wave solutions and unbounded monotonic travelling-wave solutions for equations of the form (1.1) with power-law coefficients. In particular, we shall see that all the different alternatives implicit in Theorem 11.3 are viable.

By Theorem 2.11, the partial differential equation (1.1) admits an unbounded monotonic travelling-wave solution with speed σ decreasing to 0 if and only if the integral equation (1.9) has a solution θ satisfying the integrability condition on $[0,\infty)$. Moreover, by Corollary 2.38, equation (1.1) admits an unbounded strict semi-wavefront solution with speed σ decreasing to 0 if and only if (1.9) has such a solution θ satisfying the additional constraint

$$\int_1^\infty \frac{a'(s)}{\theta(s)}\,ds < \infty. \tag{11.8}$$

Plainly, to check this criterion, it suffices to consider the maximal solution of the integral equation should the latter admit more than one solution. In this light, Theorem 11.2 is a consequence of Lemmata 2.14(i) and 2.16(i). The remaining results can be obtained as follows.

Proof of Theorem 11.3. Suppose that the set S of wave speeds for which equation (1.1) has an unbounded monotonic travelling-wave solution decreasing to 0 is not empty. Then by Theorem 2.11 and Lemma 2.18, S is an interval which is unbounded above. Define $\sigma^* := \inf S$, and, for every $\sigma \in S$ let $\theta(\cdot;\sigma)$ denote the maximal solution of equation (1.9). By Lemma 2.18(ii) there holds

$$\theta(\delta;\sigma_2) \geq \theta(\delta;\sigma_1) + (\sigma_2 - \sigma_1)\delta \quad \text{for all } \delta > 0 \quad \text{and } \sigma_2 \geq \sigma_1 \tag{11.9}$$

such that $\sigma_1 \in S$. The absurdity of the passage to the limit $\sigma_1 \to -\infty$ in this inequality demonstrates that σ^* must be finite. Furthermore this inequality shows that $\overline{\theta}(s) := \lim_{\sigma\downarrow\sigma^*} \theta(s;\sigma)$ is well-defined for all $s \geq 0$. Applying the Monotone Convergence Theorem one may subsequently pass to the limit $\sigma \downarrow \sigma^*$ in (1.9) to deduce that $\overline{\theta}$ satisfies (1.9) with $\sigma = \sigma^*$ on $[0,\infty)$. Whence, by Lemma 2.24, $\overline{\theta}$ also satisfies the integrability condition on $[0,\infty)$. This proves that $S = [\sigma^*,\infty)$. Now, let us fix $\sigma \geq \sigma^*$ and $\delta > 0$. According to Lemma 4.8, for every $0 < \varrho \leq 1$

equation (1.9) has one and only one solution $\theta(\cdot;\sigma;\varrho)$ on an interval $[0,\Delta(\varrho))$ with the following properties: $\Delta(\varrho)=\infty$ or $\theta(s;\sigma,\varrho)\to 0$ as $s\uparrow\Delta(\varrho)$, and, given any $0<\varrho_1<\varrho_2\le 1$ there holds $0<\Delta(\varrho_1)\le\Delta(\varrho_2)\le\Delta(1)=\infty$ and $\theta(s;\sigma,\varrho_1)<\theta(s;\sigma,\varrho_2)\le\theta(s;\sigma;1)=\theta(s;\sigma)$ for all $0<s<\Delta(\varrho_1)$. Let $R_0:=\{0<\varrho\le 1:\Delta(\varrho)\ge\delta\}$, $R_1:=\{\varrho\in R_0:\Delta(\varrho)=\infty\}$, $R_2:=\{\varrho\in R_0:\theta(\delta;\sigma;\varrho)\ge\theta(\delta;\sigma^*)\}$, and, $\varrho_i^*:=\inf R_i$ for $i=0,1,2$. From the proof of Lemma 4.8, it can be deduced that $0<\varrho_0^*<\varrho_1^*,\varrho_2^*\le 1$, that $\varrho_i^*\in R_i$ for $i=0,1,2$, and, that the function $\varrho\mapsto\theta(\delta;\sigma,\varrho)$ is continuous on $[\varrho_0^*,1]$. Subsequently Lemma 4.7 implies that $\varrho_1^*\le\varrho_2^*$. Meanwhile, by (11.9), $\varrho_2^*<1$ if $\sigma>\sigma^*$. Hence either $\varrho_1^*=1$, in which case $\sigma=\sigma^*$ and equation (1.9) has exactly one solution satisfying the integrability condition on $[0,\infty)$, or, $\varrho_1^*<1$, in which case there is a continuous order-preserving bijective mapping from $[\varrho^*,1]$ onto the set of such solutions. Composing this mapping with a linear mapping from $[0,1]$ onto $[\varrho^*,1]$ and recalling Lemma 2.25 shows that in the latter case equation (1.1) has a one-parameter family of solutions in the sense of Definition 11.1. □

Proof of Theorem 11.4. The comparison principle for solutions of the integral equation (1.9), Lemma 2.18, readily yields the conclusions of the theorem when hypotheses (a) or (b) hold. To obtain the conclusions of the theorem under hypothesis (c), we fix $\delta>0$ such that $\sigma_2\delta+b_2(\delta)>\sigma_1\delta+b_1(\delta)$. Subsequently, identifying the hypothesized solution of equation (1.9) for $i=1$ with $\theta(\cdot;\sigma^*)$ and solutions of (1.9) for $i=2$ with the functions $\theta(\cdot;\rho,\sigma)$, the argument yielding multiplicity of solutions for $\sigma>\sigma^*$ in the proof of Theorem 11.3 can be applied in this case too. □

Proof of Theorem 11.5. For the convection-diffusion equation (11.1) the integral equation (1.9) reduces to $\theta(s)=\sigma s+b(s)$. So for the existence of a nonnegative 'solution' on $[0,\infty)$ one plainly requires $\sigma\ge\sigma^*$, where σ^* denotes the quantity on the left-hand side of (11.2). On the other hand, for every $\sigma>\sigma^*$ such a function θ is positive on $(0,\infty)$. This yields the first assertion of the theorem. The second follows from the estimate

$$\min\{\sigma-\sigma^*,1\}\le\frac{\theta(s)}{\max\{s,b(s)\}}\le\sigma+1$$

for all $s>0$ and $\sigma>\max\{\sigma^*,0\}$, which can be obtained from the proof of Theorem 5.2. □

Proof of Theorem 11.6. It is easy to check that when $\sigma=0$ the integral equation

$$\theta(s)=\sigma s-\int_0^s\frac{c(r)a'(r)}{\theta(r)}\,dr \tag{11.10}$$

associated with the reaction-diffusion equation (11.3) admits the solution $\theta\equiv Q$ on $[0,\infty)$. Therefore, for every $\sigma>0$, this equation has a unique solution θ on $[0,\infty)$, by Lemmata 2.14(i) and 2.18(i). Moreover, $\theta\ge Q$ on $[0,\infty)$. Substituting this inequality in the right-hand side of (11.10) subsequently yields $\sigma s \le \theta(s) \le \sigma s + Q(s)$ for all $s > 0$. So, combining these estimates,

$$\min\{\sigma, 1\} \le \frac{\theta(s)}{\max\{s, Q(s)\}} \le \sigma + 1$$

for every $s > 0$. This gives the existence of a unique solution θ on $[0, \infty)$ for every $\sigma > 0$, for which (11.4) and (11.8) are equivalent. □

Proof of Theorem 11.7. Adapting the proof of Lemma 6.6, for the existence of a solution of (11.10) necessarily $\Lambda < \infty$, where Λ denotes the quantity on the left-hand side of (11.6). By the same token, for every $\sigma \ge 2\sqrt{\Lambda}$ this integral equation has a maximal solution θ on $[0, \infty)$ satisfying

$$\frac{\sigma + \sqrt{\sigma^2 - 4\Lambda}}{2} \le \frac{\theta(s)}{s} \le \sigma$$

for all $s > 0$. This yields the result. □

Under the conditions of Theorem 11.7 plus some additional regularity hypotheses, it was deduced in [114] that the necessary and sufficient criterion for the existence of an unbounded monotonic travelling-wave solution is

$$\sup_{0<s<\infty} \left\{ \frac{Q(s)}{s} \right\} < \infty, \tag{11.11}$$

where Q is given by (11.5). Likewise, the necessary and sufficient criterion for the existence of an unbounded strict semi-wavefront solution is that (11.7) and (11.11) hold. The conclusions of Theorem 11.7 are equivalent to the results in [114], representing a generalization. This can be deduced from the identity

$$\int_\varepsilon^s \frac{c(u)a'(u)}{u}\, du = \frac{1}{s}\int_\varepsilon^s c(u)a'(u)\, du + \int_\varepsilon^s \frac{1}{r^2}\int_\varepsilon^r c(u)a'(u)\, du\, dr$$

for any $0 < \varepsilon < s < \infty$. Whence, multiplying the limit as $\varepsilon \downarrow 0$ by $2/s$, we obtain

$$2\left\{\frac{1}{s}\int_0^s \frac{c(u)a'(u)}{u}\, du\right\} = \frac{Q^2(s)}{s^2} + \frac{1}{s}\int_0^s \frac{Q^2(r)}{r^2}\, dr$$

for all $0 < s < \infty$.

Chapter 12

Wavefronts and unbounded waves for power-law equations

With the porous media equation as prototype, equations of the class (1.1) with power-law coefficients have attracted much interest to date. In this chapter, we shall classify all the global monotonic travelling-wave solutions decreasing to 0 and all the unbounded monotonic semi-wavefront solutions decreasing to 0 for equations of this type. We begin with the power-law convection-diffusion equation, proceed to the power-law reaction-diffusion equation with linear convection, and, end with the full equation.

In Chapter 7 we have seen that the integral equation associated with a power-law reaction-convection-diffusion equation takes the form

$$\theta(s) = \sigma s + b_0 s^n - mc_0 \int_0^s \frac{r^{m+p-1}}{\theta(r)}\, dr \tag{12.1}$$

where $m > 0$, $n > 0$, $m+p > 0$, b_0 and c_0 are real parameters. Furthermore, when this equation has a unique solution θ, there holds

$$\theta(s) \sim \theta_0 s^{q_0} \quad \text{as } s \downarrow 0 \quad \text{for some } \theta_0 > 0 \text{ and } q_0 > 0. \tag{12.2}$$

Whereas when the integral equation admits more than one solution, its maximal solution satisfies such a relation. As it turns out, (12.2) holds for every solution θ of the integral equation (12.1) on $[0,\infty)$. Moreover, such a function satisfies

$$\theta(s) \sim \theta_1 s^{q_1} \quad \text{as } s \to \infty \quad \text{for some } \theta_1 > 0 \text{ and } q_1. \tag{12.3}$$

12.1 Convection-diffusion

The power-law convection-diffusion equation reads

$$u_t = (u^m)_{xx} + b_0(u^n)_x \tag{12.4}$$

with $m > 0$, $n > 0$ and b_0 parameters. Since for this partial differential equation, our integral equation reads simply $\theta(s) = \sigma s + b_0 s^n$, the following results are quite easy to obtain. To avoid the ambivalence with regard to the value of n caused by $b_0 = 0$, the latter parameter value will be considered exclusively in tandem with $n = 1$.

Theorem 12.1 (Wavefronts). *Let $m > 0$, $n > 0$ and b_0 be real numbers.*

(a) *If $n < 1$ and $b_0 > 0$, equation (12.4) has exactly one distinct wavefront solution decreasing to 0 for every wave speed $\sigma < 0$ and no such solution for any wave speed $\sigma \geq 0$.*

(b) *If $n > 1$ and $b_0 < 0$, equation (12.4) has exactly one distinct wavefront solution decreasing to 0 for every wave speed $\sigma > 0$ and no such solution for any wave speed $\sigma \leq 0$.*

(c) *If $n < 1$ and $b_0 < 0$, if $n = 1$, or, if $n > 1$ and $b_0 > 0$, equation (12.4) has no wavefront solutions decreasing to 0.*

Theorem 12.2 (Behaviour). *Fix $m > 0$, $n > 0$ and b_0. Let f be a wavefront solution of equation (12.4) decreasing to 0, and, q_0 and θ_0 be given by Table 12.1. Then f is a wavefront solution from $\alpha := |\sigma/b_0|^{1/(n-1)}$ to 0. There holds $f(\xi) < \alpha$ for all $-\infty < \xi < \infty$, and,*

$$\{\alpha - f(\xi)\}^{-1} f'(\xi) \to -\frac{(n-1)}{m}\sigma\alpha^{1-m} \quad \text{as } \xi \to -\infty.$$

Furthermore, if $m > q_0$, the support of f is bounded above, and,

$$f^{m-q_0-1}(\xi) f'(\xi) \to -\frac{\theta_0}{m} \tag{12.5}$$

as $\xi \uparrow \Xi_0$, where $\Xi_0 := \sup\{\xi \in (-\infty, \infty) : f(\xi) > 0\}$. Whereas, if $m \leq q_0$, the support of f is unbounded above, and, (12.5) holds as $\xi \to \infty$.

n	b_0	wave speed	q_0	θ_0	q_1	θ_1
$n < 1$	$b_0 > 0$	$\sigma > 0$	n	b_0	1	σ
		$\sigma = 0$	n	b_0	n	b_0
		$\sigma < 0$	n	b_0	none	none
$n = 1$	any b_0	$\sigma > -b_0$	1	$\sigma - b_0$	1	$\sigma - b_0$
$n > 1$	$b_0 > 0$	$\sigma > 0$	1	σ	n	b_0
		$\sigma = 0$	n	b_0	n	b_0
	$b_0 < 0$	$\sigma > 0$	1	σ	none	none

Table 12.1: Values for which a solution of (12.1) with $c_0 = 0$ satisfies (12.2) and (12.3).

Theorem 12.3 (Unbounded waves). *Let $m > 0$, $n > 0$ and b_0 be real numbers.*

(a) *If $n < 1$ and $b_0 > 0$, equation (12.4) has exactly one distinct unbounded monotonic travelling-wave solution decreasing to 0 for every wave speed $\sigma \geq 0$ and no such solution for any wave speed $\sigma < 0$. This solution is global if and only if $m \geq 1$, or, $m \geq n$ and $\sigma = 0$.*

(b) *If $n = 1$, equation (12.4) has exactly one distinct unbounded monotonic travelling-wave solution decreasing to 0 for every wave speed $\sigma > -b_0$ and no such solution for any wave speed $\sigma \leq -b_0$. This solution is global if and only if $m \geq 1$.*

(c) *If $n > 1$ and $b_0 > 0$, equation (12.4) has exactly one distinct unbounded monotonic travelling-wave solution decreasing to 0 for every wave speed $\sigma \geq 0$ and no such solution for any wave speed $\sigma < 0$. This solution is global if and only if $m \geq n$.*

(d) *If $n \neq 1$ and $b_0 < 0$, equation (12.4) has no unbounded monotonic travelling-wave solutions decreasing to 0.*

Theorem 12.4 (Behaviour). *Fix $m > 0$, $n > 0$ and b_0. Let f be an unbounded monotonic travelling-wave solution of equation (12.4) decreasing to 0, and, q_0, θ_0, q_1, and θ_1 be given by Table 12.1. Then if $q_1 \leq m$ the solution f is global and there holds*

$$f^{m-q_1-1}(\xi)f'(\xi) \to -\frac{\theta_1}{m} \tag{12.6}$$

as $\xi \to -\infty$. On the other hand, if $q_1 > m$, then f is a strict semi-wavefront and (12.6) holds as $\xi \downarrow \omega$ for some finite ω. Furthermore, if $q_0 < m$ the support of f is bounded above, and, (12.5) holds as $\xi \uparrow \Xi_0$, where Ξ_0 is the maximum of the support of f. Whereas, if $q_0 \geq m$ the support of f is unbounded above and (12.5) holds as $\xi \to \infty$.

12.2 Reaction-diffusion with linear convection

Consider the equation

$$u_t = (u^m)_{xx} + b_0 u_x + \begin{cases} c_0 u^p & \text{for } u > 0 \\ 0 & \text{for } u = 0, \end{cases} \tag{12.7}$$

and, note that equation (12.7) with $c_0 \neq 0$ admits no bounded monotonic global travelling-wave solutions decreasing to 0 whatsoever by Corollary 2.42.

Theorem 12.5 (Existence). *Let $m > 0$, $m + p > 0$, b_0 and $c_0 \neq 0$ be real numbers.*

(a) *If $c_0 < 0$, equation (12.7) has exactly one distinct unbounded monotonic travelling-wave solution decreasing to 0 for every wave speed σ. This solution is global if and only if $m \geq \max\{p, 1\}$, $m \geq p$ and $\sigma \leq -b_0$, or, $\max\{m, 1\} \geq p$ and $\sigma < -b_0$.*

(b) *If $c_0 > 0$ and $m + p = 2$, equation (12.7) has a one-parameter family of distinct unbounded monotonic travelling-wave solutions decreasing to 0 in the sense of Definition 11.1 for every wave speed $\sigma > 2\sqrt{mc_0} - b_0$, exactly one such distinct solution with wave speed $2\sqrt{mc_0} - b_0$, and, no such solution for any wave speed $\sigma < 2\sqrt{mc_0} - b_0$. These solutions are global if and only if $m \geq 1$.*

(c) *If $c_0 > 0$ and $m + p \neq 2$, equation* (12.7) *has no unbounded monotonic travelling-wave solutions decreasing to* 0.

Theorem 12.6 (Behaviour). *Fix $m > 0$, $m + p > 0$, b_0 and $c_0 \neq 0$. Let f be an unbounded monotonic travelling-wave solution of equation* (12.7) *decreasing to* 0, *and, q_0, θ_0, q_1, and θ_1 be given by Table* 12.2. *Then, verbatim, the conclusions of Theorem* 12.4 *hold. Moreover, the number of distinct travelling-wave solutions for which these conclusions apply is as stated in Table* 12.2.

In the light of Theorem 2.11 and its corollaries, to prove these theorems it is enough to confirm the following lemma.

Lemma 12.7. *Let $m > 0$, $n = 1$, $m+p > 0$, b_0 and $c_0 \neq 0$ be real numbers. Then the integral equation* (12.1) *admits a solution θ satisfying the integrability condition on $[0, \infty)$ only for those values of c_0, $m + p$ and σ shown in Table* 12.2. *Such a solution satisfies* (12.2) *and* (12.3) *with the values shown in this table. Moreover, the number of solutions satisfying these relations is as stated in Table* 12.2.

Proof. Without loss of generality we may suppose that $b_0 = 0$.

(a) Suppose that $c_0 < 0$. Then equation (12.1) has a unique solution θ on $[0, \infty)$ which is positive on $(0, \infty)$ by Lemma 2.16(i). Moreover, by the proof of Theorem 7.1, θ satisfies (12.2) for some $\theta_0 > 0$ and $q_0 > 0$, where the value of q_0 can be read from Table 7.1. The corresponding value of θ_0 may be found by substitution in (12.1). It therefore remains to examine the behaviour of $\theta(s)$ as $s \to \infty$. We distinguish six cases. Throughout we let

$$Q(s) := \sqrt{\frac{2m}{m+p}|c_0|s^{(m+p)/2}}.$$

(i) $\sigma = 0$. In this case it can be verified that $\theta \equiv Q$.

(ii) $m + p = 2$. In this case θ can also be computed explicitly as

$$\theta(s) = \frac{\sigma + \sqrt{\sigma^2 - 4mc_0}}{2}s.$$

(iii) $m + p > 2$ and $\sigma > 0$. From Lemmata 2.14(i) and 2.18(i) and the case (i) above, it follows that

$$\theta(s) \geq Q(s) \quad \text{for all } s > 0. \tag{12.8}$$

Substituting this inequality in the right-hand side of (12.1) subsequently yields

$$\theta(s) \leq \sigma s + Q(s) \quad \text{for all } s > 0. \tag{12.9}$$

These two inequalities imply $\theta(s) \sim Q(s)$ as $s \to \infty$.

c_0	$m+p$	wave speed	q_0	θ_0	q_1	θ_1	No.
$c_0<0$	$m+p>2$	$\sigma>-b_0$	1	$\sigma+b_0$	$(m+p)/2$	$\sqrt{\frac{2m}{m+p}\,\lvert c_0\rvert}$	1
		$\sigma=-b_0$	$(m+p)/2$	$\sqrt{\frac{2m}{m+p}\,\lvert c_0\rvert}$	$(m+p)/2$	$\sqrt{\frac{2m}{m+p}\,\lvert c_0\rvert}$	1
		$\sigma<-b_0$	$m+p-1$	$mc_0/(\sigma+b_0)$	$(m+p)/2$	$\sqrt{\frac{2m}{m+p}\,\lvert c_0\rvert}$	1
	$m+p=2$	all σ	1	$\frac{\sigma+b_0+\sqrt{(\sigma+b_0)^2-4mc_0}}{2}$	1	$\frac{\sigma+b_0+\sqrt{(\sigma+b_0)^2-4mc_0}}{2}$	1
	$m+p<2$	$\sigma>-b_0$	$(m+p)/2$	$\sqrt{\frac{2m}{m+p}\,\lvert c_0\rvert}$	1	$\sigma+b_0$	1
		$\sigma=-b_0$	$(m+p)/2$	$\sqrt{\frac{2m}{m+p}\,\lvert c_0\rvert}$	$(m+p)/2$	$\sqrt{\frac{2m}{m+p}\,\lvert c_0\rvert}$	1
		$\sigma<-b_0$	$(m+p)/2$	$\sqrt{\frac{2m}{m+p}\,\lvert c_0\rvert}$	$m+p-1$	$mc_0/(\sigma+b_0)$	1
$c_0>0$	$m+p=2$	$\sigma>2\sqrt{mc_0}-b_0$	1	$\frac{\sigma+b_0+\sqrt{(\sigma+b_0)^2-4mc_0}}{2}$	1	$\frac{\sigma+b_0+\sqrt{(\sigma+b_0)^2-4mc_0}}{2}$	1
			1	$\frac{\sigma+b_0-\sqrt{(\sigma+b_0)^2-4mc_0}}{2}$	1	$\frac{\sigma+b_0+\sqrt{(\sigma+b_0)^2-4mc_0}}{2}$	∞
			1	$\frac{\sigma+b_0-\sqrt{(\sigma+b_0)^2-4mc_0}}{2}$	1	$\frac{\sigma+b_0-\sqrt{(\sigma+b_0)^2-4mc_0}}{2}$	1
		$\sigma=2\sqrt{mc_0}-b_0$	1	$\sqrt{mc_0}$	1	$\sqrt{mc_0}$	1

Table 12.2: Values for which a solution of (12.1) with $n=1$ and $c_0\neq 0$ satisfies (12.2) and (12.3), and, the corresponding number of solutions.

(iv) $m+p>2$ and $\sigma<0$. Arguing analogously to the previous case, (12.8) and (12.9) hold with the inequalities reversed. This gives the identical behaviour for $s\to\infty$.

(v) $m+p<2$ and $\sigma>0$. Following the proof of case (iii), the solution θ of (12.1) again satisfies (12.9). Substituting (12.9) in the right-hand side of (12.1) then also yields

$$\theta(s)\geq\sigma s-mc_0\int_0^s\frac{r^{m+p-1}}{\sigma r+Q(r)}\,dr\qquad\text{for all }s>0.\tag{12.10}$$

Together, (12.9) and (12.10) yield $\theta(s)\sim\sigma s$ as $s\to\infty$.

(vi) $m+p<2$ and $\sigma<0$. This last case requires a little more work than the previous ones. For fixed $\varepsilon>0$ consider the function

$$\psi(s):=(1+\varepsilon)\frac{mc_0}{\sigma}s^{m+p-1}.$$

We assert that there exists an $s^*>0$ such that

$$\theta(s)\leq\psi(s)\qquad\text{for all }s\geq s^*.\tag{12.11}$$

To confirm this assertion, suppose, to start with that there exists an $s_0>0$ such that

$$\theta(s)\geq\psi(s)\qquad\text{for all }s>s_0.\tag{12.12}$$

Then by (12.1)

$$\theta(s_1)-\theta(s_0)=\sigma(s_1-s_0)-mc_0\int_{s_0}^{s_1}\frac{r^{m+p-1}}{\theta(r)}\,dr\tag{12.13}$$

for all $s_1>s_0$. Inserting (12.12) subsequently gives

$$(1+\varepsilon)\frac{mc_0}{\sigma}s_1^{m+p-1}-\theta(s_0)<\frac{\varepsilon}{1+\varepsilon}\sigma(s_1-s_0).$$

However, dividing by s_1, this is nonsensical in the limit $s_1\to\infty$. So, if our assertion is not true, there exist arbitrarily large s_0 and $s_1>s_0$ such that

$$\theta(s_i)=\psi(s_i)\qquad\text{for }i=0,1\tag{12.14}$$

and

$$\theta(s)>\psi(s)\qquad\text{for }s_0<s<s_1.\tag{12.15}$$

However, using (12.14) and (12.15) to eliminate θ from (12.13), this implies

$$(1+\varepsilon)\frac{mc_0}{\sigma}\left(s_1^{m+p-1}-s_0^{m+p-1}\right)<\frac{\varepsilon}{1+\varepsilon}\sigma\left(s_1-s_0\right).$$

Whence, multiplying by $(1+\varepsilon)\sigma$ and using the Mean Value Theorem, we obtain

$$\varepsilon\sigma^2 < (1+\varepsilon)^2 m\,|(m+p-1)c_0|\,s_0^{m+p-2}.$$

This contradicts the supposition that s_0 can be arbitrarily large. Therefore our assertion must be true. Similarly, we can show that there is an $s^* > 0$ such that

$$\theta(s) \geq (1-\varepsilon)\frac{mc_0}{\sigma}s^{m+p-1} \quad \text{for all } s \geq s^*. \tag{12.16}$$

The inequalities (12.11) and (12.16) give the required behaviour.

(b) Suppose that $c_0 > 0$ and $m+p=2$. Then equation (12.1) has no solution for $\sigma < 2\sqrt{mc_0}$ by Lemma 2.15(ii); while if $\sigma \geq 2\sqrt{mc_0}$ the results are given explicitly by Lemma 2.19.

(c) Suppose that $c_0 > 0$ and $m+p \neq 2$. In this case, from the proof of Theorem 7.1 it follows that equation (12.1) has no solution if $m+p<2$, or, if $m+p>2$ and $\sigma \leq 0$. The outstanding task is therefore to demonstrate that equation (12.1) has no solution on $[0,\infty)$ when $m+p>2$ and $\sigma>0$. From (12.1) though, we have the estimate $\theta(s) \leq \sigma s$ for any such solution θ. Substituting this inequality in the right-hand side of (12.1) subsequently implies

$$\theta(s) \leq s\left\{\sigma - \frac{mc_0}{(m+p-1)\sigma}s^{m+p-2}\right\}.$$

Whence we have a contradiction for large s. □

12.3 Reaction-convection-diffusion

For the full power-law reaction-convection-diffusion equation

$$u_t = (u^m)_{xx} + b_0(u^n)_x + \begin{cases} c_0u^p & \text{for } u>0 \\ 0 & \text{for } u=0 \end{cases} \tag{12.17}$$

with $b_0 \neq 0$ and $c_0 \neq 0$, the following theorems give a characterization of unbounded monotonic travelling-wave solutions decreasing to 0 extending those in the preceding sections.

Theorem 12.8. *Let $m>0$, $0<n<1$ or $n>1$, $m+p>0$, $b_0 \neq 0$ and $c_0 \neq 0$ be real numbers.*

(a) *If $c_0 < 0$, equation (12.17) has exactly one distinct unbounded monotonic travelling-wave solution decreasing to 0 for every wave speed σ. When $b_0 > 0$ this solution is global if and only if $m \geq \max\{n,p,1\}$, $m \geq \max\{n,p\}$ and $\sigma \leq 0$, or, $\max\{m,1\} \geq \max\{n,p\}$ and $\sigma < 0$. When $b_0 < 0$ this solution is global if and only if $\max\{m,n\} \geq \max\{p,1\}$, $\max\{m,n\} \geq p$ and $\sigma \leq 0$, or, $\max\{m,n,1\} \geq p$ and $\sigma < 0$.*

(b) *If* $c_0 > 0$, $m + p = 2$ *and* $b_0 > 0$, *equation* (12.17) *has a one-parameter family of distinct unbounded monotonic travelling-wave solutions decreasing to* 0 *in the sense of Definition* 11.1 *for every wave speed* $\sigma \geq 2\sqrt{mc_0}$ *and no such solution for any wave speed* $\sigma < 2\sqrt{mc_0}$. *All these solutions are global if* $m \geq \max\{n, 1\}$, *precisely one solution with every given wave speed is global if* $n > m \geq 2 - n$, *and, none are global if* $m < \min\{2 - n, 1\}$.

(c) *If* $c_0 > 0$, $2\max\{n, 1\} > m + p > 2\min\{n, 1\}$ *and* $b_0 > 0$, *there exists a* $\sigma^* > 0$ *such that equation* (12.17) *has a one-parameter family of distinct unbounded monotonic travelling-wave solutions decreasing to* 0 *for every wave speed* $\sigma > \sigma^*$, *exactly one such solution with wave speed* $\sigma = \sigma^*$, *and, no such solution for any wave speed* $\sigma < \sigma^*$. *All these solutions are global if* $m \geq \max\{n, 1\}$, *precisely one solution with every given wave speed is global if* $m < \max\{n, 1\}$ *and* $p \leq \max\{n, 1\}$, *and, none are global if* $p > \max\{n, 1\}$.

(d) *If* $c_0 > 0$, $m + p = 2n$ *and* $b_0 > 2\sqrt{mc_0/n}$, *equation* (12.17) *has a one-parameter family of distinct unbounded monotonic travelling-wave solutions decreasing to* 0 *for every wave speed* $\sigma \geq 0$ *and no such solution for any wave speed* $\sigma < 0$. *All these solutions are global if* $m \geq \max\{n, 1\}$, *precisely one solution with every given wave speed* $\sigma > 0$ *is global and all the solutions with wave speed* 0 *are global if* $1 > m \geq n$, *precisely one solution with every given wave speed* $\sigma > 0$ *is global and none of the solutions with wave speed* 0 *are global if* $n > m \geq 2n - 1$, *and, none at all of the solutions are global if* $m < \min\{2n - 1, n\}$.

(e) *If* $c_0 > 0$, $m + p = 2n$ *and* $b_0 = 2\sqrt{mc_0/n}$, *equation* (12.17) *has a one-parameter family of distinct unbounded monotonic travelling-wave solutions decreasing to* 0 *for every wave speed* $\sigma > 0$, *exactly one such solution with wave speed* $\sigma = 0$, *and no such solution for any wave speed* $\sigma < 0$. *All these solutions are global if* $m \geq \max\{n, 1\}$, *precisely one solution with every given wave speed* $\sigma \geq 0$ *is global if* $1 > m \geq n$, *precisely one solution with every given wave speed* $\sigma > 0$ *is global and the solution with wave speed* 0 *is not global if* $n > m \geq 2n - 1$, *and, none at all of the solutions are global if* $m < \min\{2n - 1, n\}$.

(f) *If* $c_0 > 0$ *and* $m+p > 2\max\{n, 1\}$, *if* $c_0 > 0$ *and* $m+p < 2\min\{n, 1\}$, *if* $c_0 > 0$ *and* $b_0 < 0$, *or, if* $c_0 > 0$, $m + p = 2n$ *and* $b_0 < 2\sqrt{mc_0/n}$, *equation* (12.17) *has no unbounded monotonic travelling-wave solutions decreasing to* 0.

Theorem 12.9 (Behaviour). *Fix* $m > 0$, $0 < n < 1$ *or* $n > 1$, $m + p > 0$, $b_0 \neq 0$ *and* $c_0 \neq 0$. *Let* f *be an unbounded monotonic travelling-wave solution of equation* (12.17) *decreasing to* 0. *Also, let* q_0, θ_0, q_1, *and* θ_1 *be given by Table* 12.3, 12.4, 12.5 *or* 12.6 *according to whether* $c_0 < 0$ *and* $n < 1$, $c_0 < 0$ *and* $n > 1$, $c_0 > 0$ *and* $n < 1$, *or,* $c_0 > 0$ *and* $n > 1$ *respectively. Then, verbatim, the conclusions of Theorem* 12.4 *hold. Moreover, when* $c_0 > 0$ *the number of distinct travelling-wave solutions for which these conclusions apply is as stated in Table* 12.5 *for* $n < 1$ *and Table* 12.6 *for* $n > 1$.

$m+p$	b_0	wave speed	q_0	θ_0	q_1	θ_1
$m+p>2$	$b_0>0$	all σ	n	b_0	$(m+p)/2$	$\sqrt{\frac{2m}{m+p}\lvert c_0\rvert}$
	$b_0<0$	all σ	$m+p-n$	mc_0/nb_0	$(m+p)/2$	$\sqrt{\frac{2m}{m+p}\lvert c_0\rvert}$
$m+p=2$	$b_0>0$	all σ	n	b_0	1	$\frac{\sigma+\sqrt{\sigma^2-4mc_0}}{2}$
	$b_0<0$	all σ	$2-n$	mc_0/nb_0	1	$\frac{\sigma+\sqrt{\sigma^2-4mc_0}}{2}$
$2>m+p>2n$	$b_0>0$	$\sigma>0$	n	b_0	1	σ
		$\sigma=0$	n	b_0	$(m+p)/2$	$\sqrt{\frac{2m}{m+p}\lvert c_0\rvert}$
		$\sigma<0$	n	b_0	$m+p-1$	mc_0/σ
	$b_0<0$	$\sigma>0$	$m+p-n$	mc_0/nb_0	1	σ
		$\sigma=0$	$m+p-n$	mc_0/nb_0	$(m+p)/2$	$\sqrt{\frac{2m}{m+p}\lvert c_0\rvert}$
		$\sigma<0$	$m+p-n$	mc_0/nb_0	$m+p-1$	mc_0/σ
$m+p=2n$	$b_0\neq 0$	$\sigma>0$	n	$\frac{b_0+\sqrt{b_0^2-4mc_0/n}}{2}$	1	σ
		$\sigma=0$	n	$\frac{b_0+\sqrt{b_0^2-4mc_0/n}}{2}$	n	$\frac{b_0+\sqrt{b_0^2-4mc_0/n}}{2}$
		$\sigma<0$	n	$\frac{b_0+\sqrt{b_0^2-4mc_0/n}}{2}$	$2n-1$	mc_0/σ
$m+p<2n$	$b_0>0$	$\sigma>0$	$(m+p)/2$	$\sqrt{\frac{2m}{m+p}\lvert c_0\rvert}$	1	σ
		$\sigma=0$	$(m+p)/2$	$\sqrt{\frac{2m}{m+p}\lvert c_0\rvert}$	n	b_0
		$\sigma<0$	$(m+p)/2$	$\sqrt{\frac{2m}{m+p}\lvert c_0\rvert}$	$m+p-1$	mc_0/σ
	$b_0<0$	$\sigma>0$	$(m+p)/2$	$\sqrt{\frac{2m}{m+p}\lvert c_0\rvert}$	1	σ
		$\sigma=0$	$(m+p)/2$	$\sqrt{\frac{2m}{m+p}\lvert c_0\rvert}$	$m+p-n$	mc_0/nb_0
		$\sigma<0$	$(m+p)/2$	$\sqrt{\frac{2m}{m+p}\lvert c_0\rvert}$	$m+p-1$	mc_0/σ

Table 12.3: Values for which the solution of (12.1) with $n<1$, $b_0\neq 0$ and $c_0<0$ satisfies (12.2) and (12.3).

$m+p$	b_0	wave speed	q_0	θ_0	q_1	θ_1
$m+p>2n$	$b_0>0$	$\sigma>0$	1	σ	$(m+p)/2$	$\sqrt{\frac{2m}{m+p}}\,\lvert c_0\rvert$
		$\sigma=0$	n	b_0	$(m+p)/2$	$\sqrt{\frac{2m}{m+p}}\,\lvert c_0\rvert$
		$\sigma<0$	$m+p-1$	mc_0/σ	$(m+p)/2$	$\sqrt{\frac{2m}{m+p}}\,\lvert c_0\rvert$
	$b_0<0$	$\sigma>0$	1	σ	$(m+p)/2$	$\sqrt{\frac{2m}{m+p}}\,\lvert c_0\rvert$
		$\sigma=0$	$m+p-n$	mc_0/nb_0	$(m+p)/2$	$\sqrt{\frac{2m}{m+p}}\,\lvert c_0\rvert$
		$\sigma<0$	$m+p-1$	mc_0/σ	$(m+p)/2$	$\sqrt{\frac{2m}{m+p}}\,\lvert c_0\rvert$
$m+p=2n$	$b_0\neq 0$	$\sigma>0$	1	σ	n	$\frac{b_0+\sqrt{b_0^2-4mc_0/n}}{2}$
		$\sigma=0$	n	$\frac{b_0+\sqrt{b_0^2-4mc_0/n}}{2}$	n	$\frac{b_0+\sqrt{b_0^2-4mc_0/n}}{2}$
		$\sigma<0$	$2n-1$	mc_0/σ	n	$\frac{b_0+\sqrt{b_0^2-4mc_0/n}}{2}$
$2n>m+p>2$	$b_0>0$	$\sigma>0$	1	σ	n	b_0
		$\sigma=0$	$(m+p)/2$	$\sqrt{\frac{2m}{m+p}}\,\lvert c_0\rvert$	n	b_0
		$\sigma<0$	$m+p-1$	mc_0/σ	n	b_0
	$b_0<0$	$\sigma>0$	1	σ	$m+p-n$	mc_0/nb_0
		$\sigma=0$	$(m+p)/2$	$\sqrt{\frac{2m}{m+p}}\,\lvert c_0\rvert$	$m+p-n$	mc_0/nb_0
		$\sigma<0$	$m+p-1$	mc_0/σ	$m+p-n$	mc_0/nb_0
$m+p=2$	$b_0>0$	all σ	1	$\frac{\sigma+\sqrt{\sigma^2-4mc_0}}{2}$	n	b_0
	$b_0<0$	all σ	1	$\frac{\sigma+\sqrt{\sigma^2-4mc_0}}{2}$	$2-n$	mc_0/nb_0
$m+p<2$	$b_0>0$	all σ	$(m+p)/2$	$\sqrt{\frac{2m}{m+p}}\,\lvert c_0\rvert$	n	b_0
	$b_0<0$	all σ	$(m+p)/2$	$\sqrt{\frac{2m}{m+p}}\,\lvert c_0\rvert$	$m+p-n$	mc_0/nb_0

Table 12.4: Values for which the solution of (12.1) with $n>1$, $b_0\neq 0$ and $c_0<0$ satisfies (12.2) and (12.3).

$m+p$	b_0	wave speed	q_0	θ_0	q_1	θ_1	No.
$m+p=2$	$b_0>0$	$\sigma>2\sqrt{mc_0}$	n	b_0	1	$\frac{\sigma+\sqrt{\sigma^2-4mc_0}}{2}$	1
			$2-n$	mc_0/nb_0	1	$\frac{\sigma+\sqrt{\sigma^2-4mc_0}}{2}$	∞
			$2-n$	mc_0/nb_0	1	$\frac{\sigma-\sqrt{\sigma^2-4mc_0}}{2}$	1
		$\sigma=2\sqrt{mc_0}$	n	b_0	1	$\sqrt{mc_0}$	1
			$2-n$	mc_0/nb_0	1	$\sqrt{mc_0}$	∞
$2>m+p>2n$	$b_0>0$	$\sigma>\sigma^*$	n	b_0	1	σ	1
			$m+p-n$	mc_0/nb_0	1	σ	∞
			$m+p-n$	mc_0/nb_0	$m+p-1$	mc_0/σ	1
		$\sigma=\sigma^*$	n	b_0	$m+p-1$	mc_0/σ	1
$m+p=2n$	$b_0>2\sqrt{mc_0/n}$	$\sigma>0$	n	$\frac{b_0+\sqrt{b_0^2-4mc_0/n}}{2}$	1	σ	1
			n	$\frac{b_0-\sqrt{b_0^2-4mc_0/n}}{2}$	1	σ	∞
			n	$\frac{b_0-\sqrt{b_0^2-4mc_0/n}}{2}$	$2n-1$	mc_0/σ	1
		$\sigma=0$	n	$\frac{b_0+\sqrt{b_0^2-4mc_0/n}}{2}$	n	$\frac{b_0+\sqrt{b_0^2-4mc_0/n}}{2}$	1
			n	$\frac{b_0-\sqrt{b_0^2-4mc_0/n}}{2}$	n	$\frac{b_0+\sqrt{b_0^2-4mc_0/n}}{2}$	∞
			n	$\frac{b_0-\sqrt{b_0^2-4mc_0/n}}{2}$	n	$\frac{b_0-\sqrt{b_0^2-4mc_0/n}}{2}$	1
	$b_0=2\sqrt{mc_0/n}$	$\sigma>0$	n	$\sqrt{mc_0/n}$	1	σ	∞
			n	$\sqrt{mc_0/n}$	$2n-1$	mc_0/σ	1
		$\sigma=0$	n	$\sqrt{mc_0/n}$	n	$\sqrt{mc_0/n}$	1

Table 12.5: Values for which a solution of (12.1) with $n<1$, $b_0\neq 0$ and $c_0>0$ satisfies (12.2) and (12.3), and, the corresponding number of solutions.

$m+p$	b_0	wave speed	q_0	θ_0	q_1	θ_1	No.
$m+p=2n$	$b_0>2\sqrt{mc_0/n}$	$\sigma>0$	1	σ	n	$\frac{b_0+\sqrt{b_0^2-4mc_0/n}}{2}$	1
			$2n-1$	mc_0/σ	n	$\frac{b_0+\sqrt{b_0^2-4mc_0/n}}{2}$	∞
			$2n-1$	mc_0/σ	n	$\frac{b_0-\sqrt{b_0^2-4mc_0/n}}{2}$	1
		$\sigma=0$	n	$\frac{b_0+\sqrt{b_0^2-4mc_0/n}}{2}$	n	$\frac{b_0+\sqrt{b_0^2-4mc_0/n}}{2}$	1
			n	$\frac{b_0-\sqrt{b_0^2-4mc_0/n}}{2}$	n	$\frac{b_0+\sqrt{b_0^2-4mc_0/n}}{2}$	∞
			n	$\frac{b_0-\sqrt{b_0^2-4mc_0/n}}{2}$	n	$\frac{b_0-\sqrt{b_0^2-4mc_0/n}}{2}$	1
	$b_0=2\sqrt{mc_0/n}$	$\sigma>0$	1	σ	n	$\sqrt{mc_0/n}$	1
			$2n-1$	mc_0/σ	n	$\sqrt{mc_0/n}$	∞
		$\sigma=0$	n	$\sqrt{mc_0/n}$	n	$\sqrt{mc_0/n}$	1
$2n>m+p>2$	$b_0>0$	$\sigma>\sigma^*$	1	σ	n	b_0	1
			$m+p-1$	mc_0/σ	n	b_0	∞
			$m+p-1$	mc_0/σ	$m+p-n$	mc_0/nb_0	1
		$\sigma=\sigma^*$	1	σ	$m+p-n$	mc_0/nb_0	1
$m+p=2$	$b_0>0$	$\sigma>2\sqrt{mc_0}$	1	$\frac{\sigma+\sqrt{\sigma^2-4mc_0}}{2}$	n	b_0	1
			1	$\frac{\sigma-\sqrt{\sigma^2-4mc_0}}{2}$	n	b_0	∞
			1	$\frac{\sigma-\sqrt{\sigma^2-4mc_0}}{2}$	$2-n$	mc_0/nb_0	1
		$\sigma=2\sqrt{mc_0}$	1	$\sqrt{mc_0}$	n	b_0	∞
			1	$\sqrt{mc_0}$	$2-n$	mc_0/nb_0	1

Table 12.6: Values for which a solution of (12.1) with $n>1$, $b_0\neq 0$ and $c_0>0$ satisfies (12.2) and (12.3), and, the corresponding number of solutions.

We recall that by Corollary 2.42, when $c_0 \neq 0$ equation (12.17) admits no bounded monotonic global travelling-wave solutions.

Theorem 12.8 is a consequence of the result for solutions of the corresponding integral equation (12.1).

Lemma 12.10. *Let the constraints of Theorem* 12.8 *hold.*

(i) *If* $c_0 < 0$, *then the integral equation* (12.1) *admits a unique solution* θ *satisfying the integrability condition on* $[0, \infty)$. *Such a solution satisfies* (12.2) *and* (12.3) *with the values shown in Table* 12.3 *for* $n < 1$ *and Table* 12.4 *for* $n > 1$.

(ii) *If* $c_0 > 0$, *then* (12.1) *admits a solution* θ *satisfying the integrability condition on* $[0, \infty)$ *only for those values of* $m + p$, b_0 *and* σ *shown in Table* 12.5 *for* $n < 1$ *and Table* 12.6 *for* $n > 1$. *Such a solution satisfies* (12.2) *and* (12.3) *with the values shown in these tables. Moreover, the number of solutions satisfying these relations is as stated in Tables* 12.5 *and* 12.6.

Many of the conclusions of this lemma and therewith Theorems 12.8 and 12.9 can be obtained from results we have already established and straightforward study of the integral equation (12.1). For instance, when $c_0 < 0$ it follows from Theorem 11.2 that equation (12.17) has exactly one distinct unbounded monotonic travelling-wave solution decreasing to 0 for every wave speed. Moreover, the behaviour of any such solution can be confirmed by proving part (i) of Lemma 12.10 along the lines of Lemma 12.7 for $c_0 < 0$. Alternatively, when $c_0 > 0$ it follows from (12.1) that any solution θ of this equation satisfies

$$\theta(s) < \sigma s + b_0 s^n \quad \text{for all } s > 0. \tag{12.18}$$

Hence if $n < 1$ we can define $L := \limsup_{s\to\infty} \theta(s)/s$ in the knowledge that $0 \leq L < \infty$. This means that given any $\varepsilon > 0$ there exists an increasing sequence $\{s_i\}_{i=0}^{\infty} \subset (0, \infty)$ such that $s_i \to \infty$ as $i \to \infty$, $(L - \varepsilon)s_i < \theta(s_i)$ for all $i \geq 0$, and, $\theta(r) < (L + \varepsilon)r$ for all $r > s_0$. Substituting these inequalities in (12.1) gives

$$\begin{aligned}(L-\varepsilon)s_i &< \sigma s_i + b_0 s_i^n - mc_0 \int_0^{s_0} \frac{r^{m+p-1}}{\theta(r)}\,dr \\ &\quad - \frac{mc_0}{(m+p-1)(L+\varepsilon)}(s_i^{m+p-1} - s_0^{m+p-1})\end{aligned}$$

for all $i \geq 0$. Dividing by s_i and passing to the limit $i \to \infty$ yields $m + p \leq 2$. Whereupon passage to the limit $\varepsilon \downarrow 0$ implies that $\sigma \geq L \geq 0$ if $m + p < 2$ and $\sigma \geq L + mc_0/L = (\sqrt{L} - \sqrt{mc_0/L})^2 + 2\sqrt{mc_0} \geq 2\sqrt{mc_0}$ if $m + p = 2$. Similarly, when $\sigma = 0$ or $n > 1$, by considering $L := \limsup_{s\to\infty} \theta(s)/s^n$ we can deduce that $m+p \leq 2n$, $b_0 \geq 0$ if $m+p < 2n$, and, $b_0 \geq 2\sqrt{mc_0/n}$ if $m+p = 2n$. Amalgamating these deductions with Theorems 2.11 and 7.1, we conclude that under the initial hypotheses of Theorem 12.8, equation (12.1) admits an unbounded monotonic travelling-wave solution decreasing to 0 only if $m+p=2n$, $b_0 \geq 2\sqrt{mc_0/n}$ and $\sigma \geq 0$; $m+p=2$, $b_0>0$ and $\sigma \geq 2\sqrt{mc_0}$; or, $2\min\{n,1\} < m+p < 2\max\{n,1\}$, $b_0>0$ and $\sigma>0$. Now, when $m+p=2n$, $b_0 = 2\sqrt{mc_0/n}$ and $\sigma=0$, or, when $m+p=2$, $b_0=0$ and $\sigma = 2\sqrt{mc_0}$, Lemma 2.19(i) states that equation (12.1) has a unique solution on $[0,\infty)$ and this solution is positive on $(0,\infty)$. Theorems 11.3 and 11.4 subsequently give all the existence results for $m+p=2n$ and $m+p=2$. The outstanding situation is that in which $2\min\{n,1\} < m+p < 2\max\{n,1\}$ and $b_0>0$, for which we already know that equation (12.1) has a solution on $[0,\infty)$ only if $\sigma>0$. In this situation, in the light of Theorems 2.11 and 11.3, to prove the existence of a number $\sigma^*>0$ such that (12.17) admits an unbounded monotonic travelling-wave solution with wave speed σ decreasing to 0 if and only if $\sigma \geq \sigma^*$, it suffices to show that (12.1) has a solution satisfying the integrability condition on $[0,\infty)$ for large σ. For this purpose, we remark to begin with, that one need only consider the case $n<1$. This follows from the observation that if θ is a solution of the integral equation (12.1) on $[0,\infty)$, so too is $s \mapsto \theta(s^{1/n})$ with σ, b_0, n, c_0 and $m+p$ replaced by b_0, σ, $1/n$, c_0/n and $(m+p)/n$ respectively. Also by rescaling the dependent and independent variables in (12.1), without loss of generality one may assume that $b_0=1$ and $c_0=1/m$. Against this background, we introduce the parameter

$$\gamma := \frac{m+p-n-1}{n-1} \tag{12.19}$$

and note that $-1<\gamma<1$. Suppose that $\gamma \leq 0$. In this case, it can be verified that $\theta_2(s) := (1+\gamma)^{-(1+\gamma)/(1-\gamma)} s^{m+p-1}$ is a solution of equation (12.1) with $c_0 = 1/m$ and $\sigma s + b_0 s^n$ replaced by $b_2(s) := \theta_2(s) + s^{m+p}/\theta_2(s)$ on $[0,\infty)$. Moreover, when $\sigma \geq (1-\gamma)(1+\gamma)^{(1+\gamma)/(1-\gamma)}$, there holds $\sigma s + s^n \geq b_2(s)$ for all $s>0$. On the other hand, supposing that $\gamma>0$, it can be verified that $\theta_2(s) := (1+\gamma)^{-1}s^n$ is similarly a solution of (12.1) with $c_0=1/m$ and $\sigma s + b_0 s^n$ replaced by $b_2(s) := \theta_2(s) + s^{m+p}/(m+p-n)\theta_2(s)$. Moreover, $\sigma s + s^n \geq b_2(s)$ for all $s>0$, when $\sigma \geq (1-\gamma)(1+\gamma)^{(1+\gamma)/(1-\gamma)}(m+p-n)^{-1/(1-\gamma)}$. Invoking Lemma 2.18(ii), this proves that for large enough σ equation (12.1) has a solution satisfying the integrability condition on $[0,\infty)$.

The exposition above provides an upper bound on the magnitude of the critical value σ^* in Theorem 12.8(c) when $\gamma \neq 0$. By further analysis of the integral equation (12.1) one can also obtain a lower bound for $\gamma \neq 0$ which is sharper than that previously obtained. We summarize these and other findings in the next theorem.

Theorem 12.11 (Critical speed estimates). *Define γ by (12.19). Then there exists a function ς, which depends only on γ and n, such that the critical wave speed σ^* described in Theorem 12.8 part (c) is given by*

$$\sigma^* = b_0^{-(1+\gamma)/(1-\gamma)}(mc_0)^{1/(1-\gamma)}\varsigma^{1/(1-\gamma)}(\gamma,n) \tag{12.20}$$

and

$$\varsigma(\gamma,1/n) = n\varsigma(-\gamma,n) \tag{12.21}$$

for all $-1 < \gamma < 1$, $0 < n < 1$ *and* $n > 1$. *There holds*

$$\varsigma(\gamma, n) \leq \frac{(1-\gamma)^{1-\gamma}(1+\gamma)^{1+\gamma}}{\max\{n, 1\} + \min\{\gamma(n-1), 0\}} \tag{12.22}$$

and

$$\varsigma(\gamma, n) \geq \frac{(1-\gamma)^{1-\gamma}(1+\gamma)^{1+\gamma}}{\max\{n, 1\} + \max\{\gamma(n-1), 0\}} \tag{12.23}$$

with strict inequality on both counts if $\gamma \neq 0$. *Furthermore, for fixed* n, b_0 *and* mc_0 *there holds*

$$\sigma^* \to 2\sqrt{mc_0} \quad \text{as } \gamma \downarrow -1,$$

$$\sigma^* \to 0 \quad \text{as } \gamma \uparrow 1 \quad \text{if } b_0 \geq 2\sqrt{mc_0/n}$$

and

$$\sigma^* \to \infty \quad \text{as } \gamma \uparrow 1 \quad \text{if } b_0 < 2\sqrt{mc_0/n}.$$

Proof. The first assertions of the theorem and (12.22) follow from the discussion above. When $\gamma \neq 0$ moreover, since (12.17) has precisely one unbounded monotonic travelling-wave solution decreasing to 0 with wave speed σ^*, while the comparison argument providing (12.22) falls under category (c) of Theorem 11.4, the last-mentioned theorem states that this inequality must be strict. To obtain (12.23), we assume $b_0 = 1$, $n < 1$ and $c_0 = 1/m$ until mentioned otherwise. We begin with the case $\gamma < 0$. Take $\sigma = (1-\gamma)(1+\gamma)^{(1+\gamma)/(1-\gamma)}\{1-\gamma(1-n)\}^{-1/(1-\gamma)}$, and, suppose that equation (12.1) has a solution θ on $[0, \infty)$. In view of (12.18), we can define a continuous function on $(0, \infty)$ by $A(s) := \sup\{\theta(r)/r^n : 0 < r \leq s\}$. We let $v(s)$ denote the largest number r in the interval $(0, s]$ such that $A(s) = \theta(r)/r^n$ if such a number exists, and, $v(s) = 0$ otherwise. Substitution in equation (12.1) gives

$$\begin{aligned} A(s)v^n(s) &= \theta(v(s)) \\ &= \sigma v(s) + v^n(s) - \int_0^{v(s)} \frac{r^{m+p-1}}{\theta(r)}\, dr \\ &\leq \sigma v(s) + v^n(s) - \frac{1}{(m+p-n)A(s)} v^{m+p-n}(s) \end{aligned} \tag{12.24}$$

for every $s > 0$ such that $v(s) > 0$. In fact, because $\widetilde{\theta}(r) = A(s)r^n$ is not an exact solution of (12.1) on $(0, v(s)]$, we obtain (12.24) with strict inequality. Setting

$$F^{\pm}(v) = \frac{\sigma v + v^n \pm \sqrt{(\sigma v + v^n)^2 - 4v^{m+p}/(m+p-n)}}{2v^n},$$

this implies $F^-(v(s)) < A(s) < F^+(v(s))$ for all $s > 0$ such that $v(s) > 0$. However, because $F^{\pm}$ are both strictly increasing functions on $(0, \infty)$ such that $F^-(v^*) = F^+(v^*) = (1+\gamma)^{-1}$ for some number $v^* > 0$, this excludes the possibility that $A(s) = (1+\gamma)^{-1}$ for any $s > 0$. Subsequently, recalling that A is continuous on $(0, \infty)$ and $A(s) < (1+\gamma)^{-1}$ for small $s > 0$ by (12.18), we must have $A(s) < (1+\gamma)^{-1}$ for every $s > 0$. In other words, $\theta(r) < (1+\gamma)^{-1}r^n$ for all $r > 0$. Substituting this inequality in the right-hand side of (12.1) and passing to the limit $s \to \infty$ we obtain a contradiction. Thus, we conclude that for the taken value σ, equation (12.1) has no solution on $[0, \infty)$. This yields (12.23) with strictness. The proof in the case $\gamma > 0$ is similar. Supposing that (12.1) with $\sigma = (1-\gamma)(1+\gamma)^{(1+\gamma)/(1-\gamma)}$ has a solution θ on $[0, \infty)$ and examining $A(s) := \sup\{\theta(r)/r^{m+p-1} : 0 < r \leq s\}$, it can be determined that $\theta(r) < (1+\gamma)^{-(1+\gamma)/(1-\gamma)}r^{m+p-1}$ for all $r > 0$, which upon substitution in (12.1) again yields a contradiction. In the case $\gamma = 0$ we note that $\theta(s) = s^n$ is an explicit solution of equation (12.1) with $\sigma = 1$ on $[0, \infty)$. Consequentially, taking $\sigma = 1$, the above argument stagnates at the deduction that $A(s) \leq 1$ for all $s > 0$ such that $v(s) \leq 1$. Nonetheless, if we let $\theta(\cdot; \sigma)$ denote the maximal solution of equation (12.1) on $[0, \infty)$ for every $\sigma \geq \sigma^*$, this does tell us that $\sigma^* \leq 1$ and $\theta(s; 1) = s^n$ for all $s > 0$. Hence, by Lemma 2.18(ii), $\theta(r; \sigma^*) \leq r^n$ for all $r > 0$. Substituting this inequality in the right-hand side of (12.1) and passing to the limit $s \to \infty$ implies $\sigma^* = 1$. This completes the proof of (12.23) for $n < 1$. The corresponding inequality for $n > 1$ is immediate via (12.21). The remaining conclusions of the theorem can be deduced from (12.20), (12.22) and the argument used to prove the first assertion in Theorem 10.17. □

Bibliographic notes

See, for instance, [127, 171] for an impression of the literature on the power-law reaction-convection-diffusion equation prior to 1996.

Travelling-wave solutions of the power-law reaction-diffusion equation, i.e. equation (12.7) with $b_0 = 0$ have been studied in great depth by Herrero and Vázquez [161] and by de Pablo and Vázquez [228]. In the earlier study [161], equation (12.7) was analysed with $b_0 = 0$ and $c_0 < 0$ and no restriction on m and p. In the later paper, the subject was (12.7) with $b_0 = 0$, $c_0 > 0$, $m > 1$ and no restriction on p. In both of these papers travelling waves were studied by means of a detailed phase-plane analysis. Besides answering the question of the existence and uniqueness of such solutions, these analyses provided explicit results on the asymptotic behaviour of the waves. A nontrivial travelling-wave solution whose support is bounded above is referred to as a *finite travelling-wave*. The authors also adopt terminology first introduced by the second author of the present work in [181]. This is to say that if a travelling wave has speed $\sigma > 0$ it is said to be a *heating-wave*, and, if $\sigma < 0$ it is said to be a *cooling-wave*. A travelling wave with speed $\sigma = 0$ is called a *stationary-wave*. In the earlier paper [161], a strict semi-wavefront solution is referred to as a *partial-wave*.

A thorough analysis of semi-wavefront solutions decreasing to 0 for equation (12.17) with $b_0 \neq 0$ and $c_0 \neq 0$ using a sophisticated phase-plane analysis has recently been conducted by de Pablo and Sánchez [227]. Terms which are used in [227] include a *local* wave to distinguish a strict semi-wavefront solution from a global solution, a *finite* wave for a travelling-wave solution whose support is bounded above or below, and, a *positive* wave for a global travelling-wave solution whose support is unbounded above and below. Also for the reader referring to [227], at the risk of causing confusion where we intend to clarify, we mention that a *bounded* wave in that article is synonymous with a global travelling-wave solution, while an *unbounded* wave is an unbounded strict semi-wavefront solution in our terminology. The present results using the integral equation approach have relied heavily on ideas in [227], notwithstanding that the integral equation approach has led to estimates of the critical wave speed σ^* that are sharper and resolve an open question in [227]. Further properties of the function ς introduced in Theorem 12.11 can be found in [227]. In particular, using the theory of anomalous exponents, for fixed $0 < n < 1$ and $n > 1$, de Pablo and Sánchez have shown that it is an analytic function of γ.

Chapter 13

Explicit travelling-wave solutions

A number of explicit nontrivial monotonic travelling-wave solutions of the nonlinear reaction-convection-diffusion equation (1.1) have been discovered by various authors. It is not the intention here to provide a survey of all these. However, a few remarks on the possibilities offered by the now apparent correspondence between travelling-wave solutions of the partial differential equation (1.1) and solutions of the integral equation (1.9) are in order.

Recall that if we have a solution θ of the integral equation (1.9) satisfying the integrability condition on an interval $[0,\delta)$ with $0<\delta\leq\ell$, then setting

$$\Xi_0 := \int_0^\nu \frac{a'(r)}{\theta(r)}\,dr \tag{13.1}$$

for an arbitrary $\nu\in(0,\delta)$, a semi-wavefront solution f of the reaction-convection-diffusion equation may be defined on $(0,\infty)$ by

$$\int_{f(\xi)}^\nu \frac{a'(r)}{\theta(r)}\,dr = \min\{\xi,\Xi_0\} \quad \text{for } \xi\geq 0. \tag{13.2}$$

If in addition $\delta<\infty$, θ satisfies the integrability condition on $[0,\delta]$, and $\theta(\delta)=c(\delta)=0$, then setting

$$\Xi_1 := -\int_\nu^\delta \frac{a'(r)}{\theta(r)}\,dr \tag{13.3}$$

and

$$-\int_\nu^{f(\xi)} \frac{a'(r)}{\theta(r)}\,dr = \max\{\xi,\Xi_1\} \quad \text{for } \xi<0, \tag{13.4}$$

this extends f to a wavefront solution on $(-\infty,\infty)$. Alternatively, if $\delta=\infty$, then (13.1) – (13.4) yield an unbounded travelling-wave solution of equation (1.1) on (Ξ_1,∞).

Thus any given solution θ of the integral equation satisfying the integrability condition on an interval $[0,\delta)$ yields an explicit semi-wavefront solution of the corresponding reaction-convection-diffusion equation if the integrals in (13.2) and (13.4) can be evaluated exactly. Any given solution θ of the integral equation satisfying the integrability condition on an interval $[0,\delta]$ with $\theta(\delta)=c(\delta)=0$ yields an explicit wavefront solution from δ to 0 if the integrals can be similarly evaluated. Any given solution θ of the integral equation satisfying the integrability condition on $[0,\infty)$ yields an explicit unbounded travelling-wave solution in the same way. To generate explicit travelling-wave solutions of the partial differential equation (1.1) an obvious strategy is therefore to look for suitable explicit solutions of the integral equation (1.9).

13.1 Power-law equations

The integral equation (1.9) contains only power-law coefficients when the underlying partial differential equation is the power-law reaction-convection-diffusion equation. As a result, it is relatively easy to find explicit solutions of this partial differential equation. Below, we document those which are known. First we consider wavefront solutions of the convection-diffusion equation. Recall that wavefront solutions do not exist when the power-law equation contains a reaction term, by Corollary 2.42. Thereafter we look for unbounded solutions of the equation with linear convection. Finally we consider unbounded travelling waves of the power-law equation with nonlinear convection.

Application 13.1. *Suppose that $m > 0$, $n > 0$ and b_0 and are real constants. Then*

$$u_t = (u^m)_{xx} + b_0(u^n)_x \tag{13.5}$$

admits the following wavefront solutions from ℓ to 0 with wave speed $\sigma = -b_0\ell^{n-1}$.

(a) *If $n > 1$, $b_0 < 0$ and $m = n$:*

$$f(\xi) = \begin{cases} \ell\left[1 - \exp\left(\dfrac{n-1}{n}\,|b_0|\,\xi\right)\right]^{1/(n-1)} & \textit{for } \xi < 0 \\ 0 & \textit{for } \xi \geq 0. \end{cases}$$

(b) *If $n > 1$, $b_0 < 0$ and $m = (n+1)/2$:*

$$f(\xi) = \begin{cases} \ell\left[\tanh\left(\dfrac{n-1}{n+1} b_0 \ell^{(n-1)/2}\xi\right)\right]^{2/(n-1)} & \textit{for } \xi < 0 \\ 0 & \textit{for } \xi \geq 0. \end{cases}$$

(c) *If $n > 1$, $b_0 < 0$ and $m = 1$:*

$$\begin{aligned} f(\xi) &= \ell\left[1 + \exp\left\{(n-1)\,|b_0|\,\ell^{n-1}\xi\right\}\right]^{-1/(n-1)} \\ &= \ell\left[\frac{1}{2} - \frac{1}{2}\tanh\left(\frac{n-1}{2}\,|b_0|\,\ell^{n-1}\xi\right)\right]^{1/(n-1)}. \end{aligned}$$

(d) *If $n < 1$, $b_0 > 0$ and $m = 1$:*

$$f(\xi) = \begin{cases} \ell\left[1 - \exp\left\{(1-n)b_0\ell^{n-1}\xi\right\}\right]^{1/(1-n)} & \textit{for } \xi < 0 \\ 0 & \textit{for } \xi \geq 0. \end{cases}$$

(e) *If $n < 1$, $b_0 > 0$ and $m = (n+1)/2$:*

$$f(\xi) = \begin{cases} \ell \left[\tanh\left(-\dfrac{1-n}{1+n} b_0 \ell^{(n-1)/2}\xi\right)\right]^{2/(1-n)} & \text{for } \xi < 0 \\ 0 & \text{for } \xi \geq 0. \end{cases}$$

(f) *If $n < 1$, $b_0 > 0$ and $m = n$:*

$$\begin{aligned} f(\xi) &= \ell \left[1 + \exp\left(\frac{1-n}{n} b_0 \xi\right)\right]^{-1/(1-n)} \\ &= \ell \left[\frac{1}{2} - \frac{1}{2}\tanh\left(\frac{1-n}{2n} b_0 \xi\right)\right]^{1/(1-n)}. \end{aligned}$$

Proof. By Theorems 9.1 and 12.1, equation (13.5) admits a wavefront solution from ℓ to 0 if and only if $n > 1$ and $b_0 < 0$, or, $n < 1$ and $b_0 > 0$. The wave speed $\sigma = -b_0 \ell^{n-1}$, and the corresponding 'solution' of the integral equation (1.9) is given quite simply by $\theta(s) = -b_0 s(\ell^{n-1} - s^{n-1})$. Subsequently the integrals in (13.2) and (13.4) are equivalent to

$$\frac{m}{b_0} \int_\nu^{f(\xi)} \frac{r^{m-2}}{\ell^{n-1} - r^{n-1}}\, dr = \min\{\xi, \Xi_0\},$$

with Ξ_0 defined in accordance, and $\Xi_1 = -\infty$. The key to the example is the observation that if $m = n$, $m = (n+1)/2$ or $m = 1$, the above integral can be evaluated explicitly. The given forms are obtained after a suitable translation of the wavefront. Note that in cases (a), (b), (d) and (e), the support of the wavefront solution is bounded above, while in cases (c) and (f) it is not, as predicted by Theorem 12.2. Moreover, the behaviour of the solutions as ξ approaches the upper or lower bounds of its support matches that reported in that theorem. □

Application 13.2. *Suppose that $m > 0$, $m + p > 0$, b_0 and c_0 are real constants. Then*

$$u_t = (u^m)_{xx} + b_0 u_x + \begin{cases} c_0 u^p & \text{for } u > 0 \\ 0 & \text{for } u = 0 \end{cases}$$

admits the following unbounded travelling-wave solutions with the stated wave speed(s).

(a) *If $c_0 < 0$, $m + p \neq 2$ and $m > p$: for $\sigma = -b_0$ by*

$$f(\xi) = \begin{cases} \left[-(m-p)\sqrt{\dfrac{|c_0|}{2m(m+p)}}\,\xi\right]^{2/(m-p)} & \text{for } \xi < 0 \\ 0 & \text{for } \xi \geq 0. \end{cases}$$

(b) *If $c_0 < 0$, $m + p \neq 2$ and $m = p$: for $\sigma = -b_0$ by*

$$f(\xi) = \exp\left(-\frac{\sqrt{|c_0|}}{p}\xi\right).$$

(c) *If $c_0 < 0$, $m + p \neq 2$ and $m < p$: for $\sigma = -b_0$ by*

$$f(\xi) = \left[(p-m)\sqrt{\frac{|c_0|}{2m(m+p)}}\xi\right]^{-2/(p-m)} \qquad \text{for } \xi > 0.$$

(d) *If $c_0 < 0$, $m + p = 2$ and $m > 1$: for all σ by*

$$f(\xi) = \begin{cases} \left[-\frac{m-1}{m}\theta_0\xi\right]^{1/(m-1)} & \text{for } \xi < 0 \\ 0 & \text{for } \xi \geq 0, \end{cases} \tag{13.6}$$

where

$$\theta_0 := \frac{\sigma + b_0 + \sqrt{(\sigma + b_0)^2 + 4m\,|c_0|}}{2}. \tag{13.7}$$

(e) *If $c_0 < 0$, $m + p = 2$ and $m = 1$: for all σ by*

$$f(\xi) = \exp\left(-\theta_0\xi\right), \tag{13.8}$$

where θ_0 is given by (13.7).

(f) *If $c_0 < 0$, $m + p = 2$ and $m < 1$: for all σ by*

$$f(\xi) = \left[\frac{1-m}{m}\theta_0\xi\right]^{-1/(1-m)} \qquad \text{for } \xi > 0, \tag{13.9}$$

where θ_0 is given by (13.7).

(g) *If $c_0 = 0$ and $m > 1$: for all $\sigma > -b_0$ by (13.6), where $\theta_0 := \sigma + b_0$.*

(h) *If $c_0 = 0$ and $m = 1$: for all $\sigma > -b_0$ by (13.8), where $\theta_0 := \sigma + b_0$.*

(i) *If $c_0 = 0$ and $m < 1$: for all $\sigma > -b_0$ by (13.9), where $\theta_0 := \sigma + b_0$.*

(j) *If $c_0 > 0$, $m + p = 2$ and $m > 1$: for all $\sigma \geq -b_0 + 2\sqrt{mc_0}$ by (13.6), where*

$$\theta_0 := \frac{\sigma + b_0 \pm \sqrt{(\sigma + b_0)^2 - 4mc_0}}{2}. \tag{13.10}$$

(k) *If $c_0 > 0$, $m + p = 2$ and $m = 1$: for all $\sigma \geq -b_0 + 2\sqrt{c_0}$ by (13.8), where θ_0 is given by (13.10).*

(l) *If* $c_0 > 0$, $m+p=2$ *and* $m<1$*: for all* $\sigma \geq -b_0 + 2\sqrt{mc_0}$ *by* (13.9)*, where* θ_0 *is given by* (13.10).

Note that the last three cases of this example provide two explicit travelling-wave solutions for each wave speed $\sigma > -b_0 + 2\sqrt{mc_0}$.

Proof. In the first three cases, the integral equation (1.9) reduces to

$$\theta(s) = m\,|c_0| \int_0^s \frac{r^{m+p-1}}{\theta(r)}\, dr.$$

It is easily verifiable that $\theta(s) = \theta_0 s^{(m+p)/2}$, where $\theta_0 := \sqrt{2m\,|c_0|\,/(m+p)}$, satisfies the equation in this instance, from which the results follow. In the remaining cases, the integral equation effectively reads

$$\theta(s) = (\sigma + b_0)s - mc_0 \int_0^s \frac{r}{\theta(r)}\, dr.$$

It can be checked that when θ_0 is as stated, $\theta(s) = \theta_0 s$ solves this equation. The explicit solutions follow. □

Application 13.3. *Suppose that* $m>0$, $n>1$ *or* $0<n<1$, $m+p>0$, $b_0 \neq 0$ *and* c_0 *are real constants. Then*

$$u_t = (u^m)_{xx} + b_0(u^n)_x + \begin{cases} c_0 u^p & \text{for } u > 0 \\ 0 & \text{for } u = 0 \end{cases}$$

admits the following unbounded travelling-wave solutions with the stated wave speed(s).

(a) *If* $c_0 < 0$, $m+p=2n$ *and* $m>n$*: for* $\sigma = 0$ *by*

$$f(\xi) = \begin{cases} \left[-\dfrac{m-n}{m}\theta_0 \xi\right]^{1/(m-n)} & \text{for } \xi < 0 \\ 0 & \text{for } \xi \geq 0, \end{cases} \tag{13.11}$$

where

$$\theta_0 := \frac{b_0 + \sqrt{b_0^2 + 4m\,|c_0|\,/n}}{2}. \tag{13.12}$$

(b) *If* $c_0 < 0$, $m+p=2n$ *and* $m=n$*: for* $\sigma = 0$ *by*

$$f(\xi) = \exp\left(-\frac{1}{n}\theta_0 \xi\right), \tag{13.13}$$

where θ_0 *is given by* (13.12).

(c) *If* $c_0 < 0$, $m+p = 2n$ *and* $m < n$: *for* $\sigma = 0$ *by*

$$f(\xi) = \left[\frac{n-m}{m}\theta_0\xi\right]^{-1/(n-m)} \qquad \text{for } \xi > 0, \tag{13.14}$$

where θ_0 *is given by* (13.12).

(d) *If* $c_0 < 0$, $m+p = n+1$, $b_0 > 0$ *and* $m > n$: *for* $\sigma = mc_0/b_0$ *by* (13.11) *where* $\theta_0 := b_0$.

(e) *If* $c_0 < 0$, $m+p = n+1$, $b_0 > 0$ *and* $m = n$: *for* $\sigma = nc_0/b_0$ *by* (13.13) *where* $\theta_0 := b_0$.

(f) *If* $c_0 < 0$, $m+p = n+1$, $b_0 > 0$ *and* $m < n$: *for* $\sigma = mc_0/b_0$ *by* (13.14) *where* $\theta_0 := b_0$.

(g) *If* $c_0 < 0$, $m+p = n+1$, $b_0 < 0$ *and* $m > 1$: *for* $\sigma = mc_0/nb_0$ *by*

$$f(\xi) = \begin{cases} \left[-\frac{m-1}{m}\sigma\xi\right]^{1/(m-1)} & \text{for } \xi < 0 \\ 0 & \text{for } \xi \geq 0. \end{cases} \tag{13.15}$$

(h) *If* $c_0 < 0$, $m+p = n+1$, $b_0 < 0$ *and* $m = 1$: *for* $\sigma = c_0/nb_0$ *by*

$$f(\xi) = \exp(-\sigma\xi). \tag{13.16}$$

(i) *If* $c_0 < 0$, $m+p = n+1$, $b_0 < 0$ *and* $m < 1$: *for* $\sigma = mc_0/nb_0$ *by*

$$f(\xi) = \left[\frac{1-m}{m}\sigma\xi\right]^{-1/(1-m)} \qquad \text{for } \xi > 0. \tag{13.17}$$

(j) *If* $c_0 = 0$, $b_0 > 0$ *and* $m > n$: *for* $\sigma = 0$ *by* (13.11) *where* $\theta_0 := b_0$.

(k) *If* $c_0 = 0$, $b_0 > 0$ *and* $m = n$: *for* $\sigma = 0$ *by* (13.13) *where* $\theta_0 := b_0$.

(l) *If* $c_0 = 0$, $b_0 > 0$ *and* $m < n$: *for* $\sigma = 0$ *by* (13.14) *where* $\theta_0 := b_0$.

(m) *If* $c_0 > 0$, $m+p = 2n$, $b_0 \geq 2\sqrt{mc_0/n}$ *and* $m > n$: *for* $\sigma = 0$ *by* (13.11) *where*

$$\theta_0 := \frac{b_0 \pm \sqrt{b_0^2 - 4mc_0/n}}{2}. \tag{13.18}$$

(n) *If* $c_0 > 0$, $m+p = 2n$, $b_0 \geq 2\sqrt{mc_0/n}$ *and* $m = n$: *for* $\sigma = 0$ *by* (13.13) *where* θ_0 *is given by* (13.18).

(o) *If* $c_0 > 0$, $m+p = 2n$, $b_0 \geq 2\sqrt{mc_0/n}$ *and* $m < n$: *for* $\sigma = 0$ *by* (13.14) *where* θ_0 *is given by* (13.18).

(p) *If* $c_0 > 0$, $m + p = n + 1$, $n > 1$, $b_0 > 0$ *and* $m > n$: *for* $\sigma = mc_0/nb_0$ *by* (13.15); *and for* $\sigma = mc_0/b_0$ *by* (13.11) *with* $\theta_0 := b_0$.

(q) *If* $c_0 > 0$, $m+p = n+1$, $n > 1$, $b_0 > 0$ *and* $m = n$: *for* $\sigma = c_0/b_0$ *by* (13.15); *and for* $\sigma = nc_0/b_0$ *by* (13.13) *with* $\theta_0 := b_0$.

(r) *If* $c_0 > 0$, $m + p = n + 1$, $n > 1$, $b_0 > 0$ *and* $1 < m < n$: *for* $\sigma = mc_0/nb_0$ *by* (13.15); *and for* $\sigma = mc_0/b_0$ *by* (13.14) *with* $\theta_0 := b_0$.

(s) *If* $c_0 > 0$, $m + p = n + 1$, $n > 1$, $b_0 > 0$ *and* $m = 1$: *for* $\sigma = c_0/nb_0$ *by* (13.16); *and for* $\sigma = c_0/b_0$ *by* (13.14) *with* $\theta_0 := b_0$.

(t) *If* $c_0 > 0$, $m + p = n + 1$, $n > 1$, $b_0 > 0$ *and* $m < 1$: *for* $\sigma = mc_0/nb_0$ *by* (13.17); *and for* $\sigma = mc_0/b_0$ *by* (13.14) *with* $\theta_0 := b_0$.

(u) *If* $c_0 > 0$, $m + p = n + 1$, $n < 1$, $b_0 > 0$ *and* $m > 1$: *for* $\sigma = mc_0/b_0$ *by* (13.11) *with* $\theta_0 := b_0$; *and for* $\sigma = mc_0/nb_0$ *by* (13.15).

(v) *If* $c_0 > 0$, $m + p = n + 1$, $n < 1$, $b_0 > 0$ *and* $m = 1$: *for* $\sigma = c_0/b_0$ *by* (13.11) *with* $\theta_0 := b_0$; *and for* $\sigma = c_0/nb_0$ *by* (13.16).

(w) *If* $c_0 > 0$, $m + p = n + 1$, $n < 1$, $b_0 > 0$ *and* $n < m < 1$: *for* $\sigma = mc_0/b_0$ *by* (13.11) *with* $\theta_0 := b_0$; *and for* $\sigma = mc_0/nb_0$ *by* (13.17).

(x) *If* $c_0 > 0$, $m+p = n+1$, $n < 1$, $b_0 > 0$ *and* $m = n$: *for* $\sigma = nc_0/b_0$ *by* (13.13) *with* $\theta_0 := b_0$; *and for* $\sigma = c_0/b_0$ *by* (13.17).

(y) *If* $c_0 > 0$, $m + p = n + 1$, $n < 1$, $b_0 > 0$ *and* $m < n$: *for* $\sigma = mc_0/b_0$ *by* (13.14) *with* $\theta_0 := b_0$; *and for* $\sigma = mc_0/nb_0$ *by* (13.17).

Proof. For cases (a) – (c) and (j) – (o) the integral equation reads

$$\theta(s) = b_0 s^n - mc_0 \int_0^s \frac{r^{2n-1}}{\theta(r)}\, dr.$$

In this event, $\theta(s) = \theta_0 s^n$ is an explicit solution when θ_0 is chosen as indicated. In the remaining cases, the equation becomes

$$\theta(s) = \sigma s + b_0 s^n - mc_0 \int_0^s \frac{r^n}{\theta(r)}\, dr.$$

When $b_0 > 0$ and $\sigma = mc_0/b_0$, a solution of this equation is $\theta(s) = b_0 s^n$. When $c_0 < 0$ and $b_0 < 0$, or, $c_0 > 0$ and $b_0 > 0$, and, $\sigma = mc_0/nb_0$, a solution is $\theta(s) = \sigma s$. Manipulation with these solutions of the integral equation provides the explicit solutions of the power-law reaction-convection-diffusion equation. □

13.2 Generalizations of the Fisher equation

The following equation, with m, p, b_0, b_1, c_0 and c_1 real numbers, embodies the Fisher equation and a number of other well-known reaction-convection-diffusion equations:

$$u_t = (u^m)_{xx} + (u(b_0 + b_1 u^p))_x + \begin{cases} u^{2-m}(1-u^p)(c_0 + c_1 u^p) & \text{for } u > 0 \\ 0 & \text{for } u = 0. \end{cases} \tag{13.19}$$

The Fisher or logistic equation is the special case where $m = p = 1$, $b_0 = b_1 = 0$, $c_0 = 1$ and $c_1 = 0$, or, alternatively where $m = 1$, $p = 1/2$, $b_0 = b_1 = 0$ and $c_0 = c_1 = 1$. When $m = p = 1$, $b_0 = b_1 = 0$ and $c_0 = c_1 = 1$, equation (13.19) may be recognized as the Newell–Whitehead or amplitude equation. When $m = p = 1$, $b_0 = b_1 = 0$, $c_0 = 0$ and $c_1 = 1$, it is the Zeldovich equation. When $m = p = 1$, $b_0 = b_1 = 0$, $-1 < c_0 < 0$ and $c_1 = 1$, it reduces to the Nagumo or bistable equation with $\alpha = -c_0$. Incidentally, the Burgers equation, the foam drainage equation, and the porous media equation with convection may also be regarded as special cases of (13.19) with $b_0 = c_0 = c_1 = 0$.

Many explicit wavefront solutions of equations of the type (13.19) with various parameter values have been found in the past, some of these being rediscovered on a number of occasions. In the light of the integral equation, all of these wavefront solutions may be considered as possessing the same generic nature and can be obtained in a unified way. We present this below.

Application 13.4. *Suppose that $m > 0$, $-1 < p < 0$ or $p > 0$, b_0, b_1, c_0 and c_1 are real constants. Suppose furthermore that*

$$b_1^2 + 4mc_1/(p+1) \geq 0 \quad \text{and} \quad p\theta_0 > 0, \tag{13.20}$$

where

$$\theta_0 := \frac{-b_1 \pm \sqrt{b_1^2 + 4mc_1/(p+1)}}{2}. \tag{13.21}$$

Then equation (13.19) admits a wavefront solution from 1 to 0 with wave speed

$$\sigma = \theta_0 + \frac{mc_0}{\theta_0} - b_0. \tag{13.22}$$

Moreover, in the following cases, such a wavefront solution is given explicitly as indicated.

(a) *If $p > 0$ and $m = p + 1$:*

$$f(\xi) = \begin{cases} \left[1 - \exp\left(\frac{p}{p+1}\theta_0 \xi\right)\right]^{1/p} & \text{for } \xi < 0 \\ 0 & \text{for } \xi \geq 0. \end{cases}$$

(b) *If $p > 0$ and $m = p/2 + 1$:*

$$f(\xi) = \begin{cases} \left[\tanh\left(-\frac{p}{p+2}\theta_0 \xi\right)\right]^{2/p} & \text{for } \xi < 0 \\ 0 & \text{for } \xi \geq 0. \end{cases}$$

(c) *If $p > 0$ and $m = 1$:*

$$\begin{aligned} f(\xi) &= [1 + \exp(p\theta_0\xi)]^{-1/p} \\ &= \left[\frac{1}{2} - \frac{1}{2}\tanh\left(\frac{p}{2}\theta_0\xi\right)\right]^{1/p}. \end{aligned}$$

(d) *If $p < 0$ and $m = 1$:*

$$f(\xi) = \begin{cases} [1 - \exp(p\theta_0\xi)]^{1/|p|} & \textit{for } \xi < 0 \\ 0 & \textit{for } \xi \geq 0. \end{cases}$$

(e) *If $p < 0$ and $m = p/2 + 1$:*

$$f(\xi) = \begin{cases} \left[\tanh\left(-\frac{p}{p+2}\theta_0\xi\right)\right]^{2/|p|} & \textit{for } \xi < 0 \\ 0 & \textit{for } \xi \geq 0. \end{cases}$$

(f) *If $p < 0$ and $m = p + 1$:*

$$\begin{aligned} f(\xi) &= \left[1 + \exp\left(\frac{p}{p+1}\theta_0\xi\right)\right]^{-1/|p|} \\ &= \left[\frac{1}{2} - \frac{1}{2}\tanh\left(\frac{p}{2(p+1)}\theta_0\xi\right)\right]^{1/|p|}. \end{aligned}$$

For $p > 0$, this gives one distinct explicit wavefront solution of (13.19) *whenever $c_1 > 0$, $c_1 = 0$ and $b_1 < 0$, or, $c_1 < 0$ and $b_1 = -2\sqrt{m\,|c_1|\,/(p+1)}$, and, two such solutions whenever $c_1 < 0$ and $b_1 < -2\sqrt{m\,|c_1|\,/(p+1)}$. Moreover, if $c_1 = -(p+1)c_0 < 0$ and $b_1 < -2\sqrt{m\,|c_1|\,/(p+1)}$, then the two solutions possess the selfsame wave speed, viz. $\sigma = -b_0 - b_1$. For $p < 0$, we have the same conclusions with b_1 replaced by $-b_1$.*

In the event that $p > 0$, when $c_0 < 0$ and $c_0 + c_1 > 0$, or $c_0 = c_1 = 0$, (13.22) is necessarily the unique wave speed for which equation (13.19) admits a wavefront solution from 1 to 0, by Theorem 8.3 part (vii). Similarly, in the event that $p < 0$, when $c_1 < 0$ and $c_0 + c_1 > 0$, or $c_0 = c_1 = 0$, this is true. Moreover, when $p > 0$, and, $c_0 \leq 0$ or $c_0 + c_1 \geq 0$, the wavefront with the wave speed (13.22) is, modulo translation, necessarily the only wavefront solution with this wave speed, by Theorem 8.7. The analogous conclusion holds for $p < 0$, and, $c_1 \leq 0$ or $c_0 + c_1 \geq 0$.

Proof. For equation (13.19) the integral equation (1.9) reads

$$\theta(s) = \sigma s + s(b_0 + b_1 s^p) - m\int_0^s \frac{r(1 - r^p)(c_0 + c_1 r^p)}{\theta(r)}\,dr.$$

By substitution it can be verified that $\theta(s) = \theta_0 s(1 - s^p)$ satisfies this equation if and only if θ_0 is given by (13.21) and σ by (13.22). Furthermore, this function will satisfy the integrability condition on $[0,1]$ if and only if it is positive on $(0,1)$. Together, this gives the necessity and sufficiency of (13.20) for θ to correspond to a wavefront solution of (13.19) from 1 to 0 with wave speed (13.22). The explicit solutions may subsequently be deduced analogously to those in Application 13.1. The final conclusions of the current example follow from further analysis of (13.21) and (13.22). □

For the specific examples of equation (13.19) mentioned at the start of this section, the above example gives the following.

- The Fisher equation, $u_t = u_{xx} + u(1-u)$, admits the wavefront solution

$$f(\xi) = [1 + \exp(\xi/\sqrt{6})]^{-2}. \tag{13.23}$$

  with wave speed $\sigma = 5/\sqrt{6}$.

- The Newell–Whitehead equation, $u_t = u_{xx} + u(1-u^2)$, admits the wavefront solution

$$f(\xi) = [1 + \exp(\xi/\sqrt{2})]^{-1} \tag{13.24}$$

  with wave speed $\sigma = 3/\sqrt{2}$, the wavefront solution $f(\xi) = -\tanh(\xi/\sqrt{2})$ with wave speed $\sigma = 0$, and the wavefront solution given by $f(\xi) = -[1 + \exp(-\xi/\sqrt{2})]^{-1}$ with wave speed $\sigma = -3/\sqrt{2}$.

- The Zeldovich equation, $u_t = u_{xx} + u^2(1-u)$, admits the wavefront solution (13.24) with wave speed $\sigma = 1/\sqrt{2}$.

- The Nagumo equation, $u_t = u_{xx} + u(1-u)(u-\alpha)$, admits the wavefront solution (13.24) with wave speed $\sigma = (1-2\alpha)/\sqrt{2}$, the wavefront solution $f(\xi) = \alpha[1 + \exp(\alpha\xi/\sqrt{2})]^{-1}$ with wave speed $\sigma = (\alpha - 2)/\sqrt{2}$, and the wavefront solution $f(\xi) = \alpha + (1-\alpha)[1 + \exp\{(1-\alpha)\xi/\sqrt{2}\}]^{-1}$ with wave speed $\sigma = (\alpha+1)/\sqrt{2}$.

With regard to the Fisher equation, we note that the obvious set of parameters $m = p = 1$, $b_0 = b_1 = 0$, $c_0 = 1$ and $c_1 = 0$ yields a blank in Application 13.4. The solution (13.23) is found by taking $m = 1$, $p = 1/2$, $b_0 = b_1 = 0$ and $c_0 = c_1 = 1$. In this guise, the explicit solution (13.23) follows from part (c). Solution (13.24) of the Newell–Whitehead equation, the solution of the Zeldovich equation, and the first of the solutions of the Nagumo equation, follow analogously. For all three, $m = p = 1$, $b_0 = b_1 = 0$ and $c_1 = 1$. The equations differ in that $c_0 = 1$, $c_0 = 0$ and $c_0 = -\alpha$, respectively. To obtain the remaining solutions of the Newell–Whitehead equation, we look for wavefront solutions from 1 to -1 and from 0 to -1, respectively. Defining $\widetilde{u} := (u+1)/2$, yields an equation of the class (13.19) with unknown $\widetilde{u}$, $m = p = 1$, $b_0 = b_1 = 0$, $c_0 = -2$ and $c_1 = 4$. Applying Application 13.4 part (c) to this equation yields the second solution of the original equation. The third is obtained similarly via $\widetilde{u} := u + 1$ which yields an equation of the class (13.19) with $m = p = 1$, $b_0 = b_1 = 0$, $c_0 = -2$ and $c_1 = 1$. The second and third solutions of the Nagumo equation are obtained in the same way. Defining $\widetilde{u} := u/\alpha$, yields an equation of the class (13.19) with unknown $\widetilde{u}$, $m = p = 1$, $b_0 = b_1 = 0$, $c_0 = -\alpha$ and $c_1 = \alpha^2$, while setting $\widetilde{u} := (u - \alpha)/(1 - \alpha)$ yields an equation that differs only in that $c_0 = \alpha(1 - \alpha)$ and $c_1 = (1 - \alpha)^2$. Applying Application 13.4(c) to the equations in $\widetilde{u}$ and then transforming back to the original variable gives the stated solutions.

It was discovered by Ablowitz and Zeppetella [2] that besides the explicit wavefront solution (13.23), the Fisher equation admits another explicit travelling-wave solution with the selfsame wave speed $\sigma = 5/\sqrt{6}$. This is the unbounded strict semi-wavefront solution, $f(\xi) = [\exp(\xi/\sqrt{6}) - 1]^{-2}$ for $\xi > 0$. The Newell–Whitehead equation displays a similar phenomenon. Besides (13.24) it admits the travelling-wave solution $f(\xi) = [\exp(\xi/\sqrt{2}) - 1]^{-1}$ for $\xi > 0$ with wave speed $\sigma = 3/\sqrt{2}$ [65]. Such coincidences have an obvious providence, when seen in terms of the integral equation for the Fisher equation. The next example demonstrates this with a generalization. The generalization is straightforward, just to indicate what is feasible. No attempt has been made to exhaust all possible explicit travelling-wave solutions of equations of the class (13.19) decreasing to 0, nor, to document the outcome in detail. The main ideas of what has to be done should be apparent, and, we leave the task of exploiting all the possibilities for equations of the class (13.19) to the interested reader.

Application 13.5. *The equation*

$$u_t = (u^m)_{xx} + \begin{cases} u^{2-m}(1 - u^p)(1 + u^p) & \text{for } u > 0 \\ 0 & \text{for } u = 0, \end{cases} \tag{13.25}$$

where $m > 0$, and, $-1 < p < 0$ or $p > 0$, simultaneously admits a wavefront solution from 1 to 0 and an unbounded travelling-wave solution decreasing to 0 with the wave speed $\sigma = (p+2)\sqrt{m/(p+1)}$. Moreover, if $m = p+1$, $m = p/2+1$, or $m = 1$, both of these solutions may be computed explicitly.

Proof. The integral equation reads

$$\theta(s) = \sigma s - m \int_0^s \frac{r(1 - r^p)(1 + r^p)}{\theta(r)}\, dr.$$

By substitution it can be verified that the functions $\theta^{\pm}(s) = \theta_0 s(1 \pm s^p)$, where $\theta_0 := \sqrt{m/(p+1)}$, simultaneously solve this equation for the stated value of σ. The function θ^- corresponds to the wavefront solution of (13.25), θ^+ to the unbounded travelling-wave solution. □

The astute reader may notice that the case $p = 0$ is avoided in Application 13.4. This borderline case is treated below.

Application 13.6. *Consider the equation*

$$\begin{aligned} u_t &= (u^m)_{xx} + (u(b_0 + b_1 \ln u))_x \\ &+ \begin{cases} u^{2-m}(\ln u)(c_0 + c_1 \ln u) & \textit{for } u > 0 \\ 0 & \textit{for } u = 0, \end{cases} \end{aligned} \tag{13.26}$$

where $m > 0$, b_0, b_1, c_0 *and* c_1 *are real constants. Suppose that*

$$b_1^2 - 4mc_1 \geq 0 \quad \textit{and} \quad \theta_0 > 0, \tag{13.27}$$

where

$$\theta_0 := \frac{-b_1 \pm \sqrt{b_1^2 - 4mc_1}}{2}. \tag{13.28}$$

Then equation (13.26) *admits a wavefront solution from* 1 *to* 0 *with wave speed*

$$\sigma = \frac{m(c_1 - c_0)}{\theta_0} - b_0. \tag{13.29}$$

Moreover, when $m = 1$ *such a wavefront solution is given explicitly by*

$$f(\xi) = \exp\{-\exp(\theta_0 \xi)\}.$$

This gives one distinct explicit wavefront solution of equation (13.26) *whenever* $c_1 < 0$, $c_1 = 0$ *and* $b_1 < 0$, *or,* $c_1 > 0$ *and* $b_1 = -2\sqrt{mc_1}$, *and, two such solutions whenever* $c_1 > 0$ *and* $b_1 < -2\sqrt{mc_1}$. *Moreover, if* $c_1 = c_0 > 0$ *and* $b_1 < -2\sqrt{mc_1}$, *then the two solutions possess the selfsame wave speed, viz.* $\sigma = -b_0$.

Proof. The proof of this example runs along the lines of Application 13.4. The integral equation reads

$$\theta(s) = \sigma s + s(b_0 + b_1 \ln s) - m \int_0^s \frac{r(\ln r)(c_0 + c_1 \ln r)}{\theta(r)}\, dr,$$

for which it can be verified that $\theta(s) = -\theta_0 s \ln s$ is a solution satisfying the integrability condition on $[0, 1]$ when (13.27) – (13.29) hold. We omit further details. □

When $c_0 < 0$ and $c_1 < 0$, or, $c_0 = c_1 = 0$, (13.29) is the unique wave speed for which equation (13.26) admits a wavefront solution from 1 to 0. Furthermore, when $c_0 \leq 0$ or $c_1 \leq 0$, the wavefront solution with speed (13.29) is unique modulo translation.

13.3 Generating further explicit solutions

In this final section, we discuss how one may generate further explicit solutions of reaction-convection-diffusion equations of the class (1.1) using the integral equation (1.9).

In general, given any diffusion term a satisfying Hypothesis 2.1 part (i), any number σ, and, a nonnegative function θ with $\theta(0) = 0$ satisfying the integrability condition on an interval $[0, \delta)$ with $0 < \delta \leq \ell$, we can construct a reaction term c in the reaction-diffusion equation

$$u_t = (a(u))_{xx} + c(u) \tag{13.30}$$

so that θ corresponds to a semi-wavefront solution f with wave speed σ decreasing to 0. The integral equation associated with (13.30) reads

$$\theta(s) = \sigma s - \int_0^s \frac{c(r)a'(r)}{\theta(r)}\,dr.$$

So, assuming that θ is differentiable, one evidently needs

$$c(s)a'(s) = \{\sigma - \theta'(s)\}\theta(s) \quad \text{for } 0 < s < \delta. \tag{13.31}$$

This yields the desired reaction term c, after taking the precaution of setting $c(0) = 0$. Now, if, for some $\nu \in (0, \delta)$, the integrals in (13.1) and (13.2) can be evaluated exactly, the constructed reaction term c will belong to a reaction-diffusion equation, for which (13.1) and (13.2) give an explicit semi-wavefront solution f with wave speed σ decreasing to 0. It is not necessary in this argument that θ be continuously differentiable. Since, we allow reaction terms which are discontinuous, it is enough that θ' be defined at all but a countable number of points in $(0, \delta)$.

As a simple illustration of the above idea, consider the following.

Application 13.7. *The equation*

$$u_t = u_{xx} - (1+u)\ln^2(1+u),$$

admits the explicit travelling-wave solution

$$f(\xi) = \exp(\exp(-\xi)) - 1 \tag{13.32}$$

with wave speed $\sigma = 1$.

Proof. The function $\theta(s) := (1+s)\ln(1+s)$ is such that $\theta(0) = 0$ and $\theta(s) > 0$ for $s > 0$. Substituting this function in (13.31) gives $c(s)a'(s) = (1+s)\ln(1+s) \cdot \{\sigma - 1 - \ln(1+s)\}$. Choosing $a(s) := s$ and $\sigma := 1$ simplifies this expression to $c(s) = -(1+s)\ln^2(1+s)$, and this extends to $s = 0$. Moreover, (13.1) and (13.2) yield $\Xi_0 = \infty$ and $\ln\{\ln(1+r)\}|_{r=f(\xi)}^{\nu} = \xi$ respectively, for any $\nu > 0$. Choosing $\nu := e - 1$, gives (13.32). □

If, in the technique described so far, $\delta < \infty$, θ satisfies the integrability condition on $[0, \delta]$, and, $\theta(\delta) = 0$, then additionally supposing that $c(\delta) = 0$ leads to a reaction term in an equation, for which (13.1) – (13.4) define an explicit wavefront solution from δ to 0 with wave speed σ. On the other hand, if $\delta = \infty$, then (13.1) – (13.4) yield an explicit unbounded travelling-wave solution of equation (13.30) with this wave speed.

To illustrate the above, we consider three simple examples.

Application 13.8. *For any real number c_0, the equation*

$$u_t = u_{xx} + \begin{cases} u(c_0 - 2\ln u - \ln^2 u)\ln^2 u & \textit{for } u > 0 \\ 0 & \textit{for } u = 0, \end{cases}$$

admits the wavefront solution

$$f(\xi) = \begin{cases} \exp(1/\xi) & \textit{for } \xi < 0 \\ 0 & \textit{for } \xi \geq 0 \end{cases}$$

with wave speed $\sigma = c_0$.

Proof. Assume $a(u) := u$, fix σ, and, consider the function $\theta(s) := s\ln^2 s$ for $0 < s \leq 1$. This function is continuously extendible to $\theta(0) = 0$, is such that $\theta(1) = 0$, and, is positive on $(0,1)$. Substituting θ in (13.31) yields $c(s) = (\sigma - 2\ln s - \ln^2 s)s\ln^2 s$. On the other hand, (13.1) gives $\Xi_0 = (-\ln\nu)^{-1}$ and (13.3) with $\delta = 1$ gives $\Xi_1 = -\infty$ for any $0 < \nu < 1$. Hence, (13.2) and (13.4) imply $-\{\ln f(\xi)\}^{-1} = \min\{\xi - \Xi_0, 0\}$ for all ξ. Inverting to find $f(\xi)$, and translating the resulting function, produces the desired result. □

Application 13.9. *Consider the equation*

$$u_t = (u^m)_{xx} + \begin{cases} 0 & \textit{for } u = 0 \\ u^{m-p}(c_0 + c_1 u^p + c_2 u^{2p}) & \textit{for } 0 < u < 1 \\ 0 & \textit{for } u = 1, \end{cases}$$

where $m > 0$, $0 < p < 2m$, c_0, c_1 and c_2 are real numbers. Then in each of the following cases, the equation admits a wavefront solution from 1 to 0 with wave speed σ if and only if $\sigma = 0$. Moreover, modulo translation, this solution is as stated below.

(i) *If $c_0 < 0$, $c_1 = -4mc_0/(2m-p)$ and $c_2 = (2m+p)c_0/(2m-p)$:*

$$f(\xi) = \begin{cases} \tanh^{2/p}\left(\frac{p}{2m}\sqrt{\frac{c_1}{2}}\xi\right) & \textit{for } \xi < 0 \\ 0 & \textit{for } \xi \geq 0. \end{cases}$$

(ii) *If $c_0 = 0$, $c_1 < 0$ and $c_2 = -(2m+p)c_1/2m$:*

$$f(\xi) = \begin{cases} 1 & \textit{for } \xi \leq 0 \\ \operatorname{sech}^{2/p}\left(\frac{p}{2m}\sqrt{|c_1|}\xi\right) & \textit{for } \xi > 0. \end{cases}$$

(iii) *If* $c_0 < 0$, $c_1 = -2mc_0/(2m-p)$ *and* $c_2 = 0$*:*

$$f(\xi) = \begin{cases} 1 & \text{for } \xi \le -m\pi/2 \\ \left[\frac{1}{2} - \frac{1}{2}\sin\left(\frac{p}{m}\sqrt{c_1}\xi\right)\right]^{1/p} & \text{for } -m\pi/2 < \xi < m\pi/2 \\ 0 & \text{for } \xi > m\pi/2. \end{cases}$$

Proof. Suppose that $a(u) := u^m$ and $\sigma := 0$. Let $\theta(s) := \theta_0 s^q(1-s^p)^n$, where $\theta_0 > 0$, $q > 0$ and $n > 0$. There holds $\theta(0) = \theta(1) = 0$, and, $\theta(s) > 0$ for $0 < s < 1$. Substituting in (13.31) gives $c(s) = -\theta_0^2 s^{2q-m}(1-s^p)^{2n-1}\{q-(q+np)s^p\}/m$. The question is, given $\nu \in (0,1)$, can we evaluate (13.1) – (13.4) exactly? In other words, can we find an invertible explicit function Φ for which

$$\Phi(\nu) - \Phi(s) = \frac{m}{\theta_0}\int_s^\nu r^{m-q-1}(1-r^p)^{-n}\,dr$$

for every $0 < s < \nu < 1$? When $q = m - p/2$ and $n = 1$, we compute $\Phi(s) = \Phi_0 \operatorname{arctanh}(s^{p/2})$ where $\Phi_0 := 2m/p\theta_0$; when $q = m$ and $n = 1/2$, we compute $\Phi(s) = -\Phi_0 \operatorname{arctanh}((1-s^p)^{1/2})$ where $\Phi_0 := 2m/p\theta_0$; and, when $q = m - p/2$ and $n = 1/2$, we compute $\Phi(s) = -\Phi_0 \arcsin(1-2s^p)$ where $\Phi_0 := m/p\theta_0$. These give the explicit solutions in parts (i), (ii) and (iii) respectively, by choosing θ_0 appropriately. The conclusions regarding the uniqueness of the wave speed and the wavefront solution follow hereafter from Theorem 10.26, in particular, part (b) of this theorem. □

Our final simple example involves the model of molecule orientation in a ferroelectric chiral smetic-C^* liquid crystal [253, 275] previously treated in Application 10.13.

Application 13.10. *For any numbers* E *and* H *excepting* $E = H = 0$*, the equation*

$$u_t = u_{xx} + \left\{E - \left(E^2 + \frac{H^2}{4}\right)\cos u\right\}\sin u \tag{13.33}$$

admits the wavefront solution

$$f(\xi) = 2\arctan\left\{\exp\left(-\sqrt{E^2 + \frac{H^2}{4}}\xi\right)\right\}$$

with wave speed

$$\sigma = \frac{E}{\sqrt{E^2 + \frac{H^2}{4}}}.$$

Proof. Recall from the justification of Application 10.14 that the integral equation associated with (13.33) admits the explicit solution

$$\theta(s) = \sqrt{E^2 + \frac{H^2}{4}}\sin s$$

when σ is as stated. Invoking (13.1) – (13.4) gives the result. □

The above idea can be extended. Given any diffusion term a satisfying Hypothesis 2.1(i), reaction term c satisfying Hypothesis 2.1(iii), number σ, and, non-negative function θ with $\theta(0) = 0$ satisfying the integrability condition on an interval $[0, \delta)$ with $0 < \delta \leq \ell$, in principle we can construct a convection term b in an equation of the class (1.1) for which θ corresponds to a semi-wavefront solution with wave speed σ decreasing to 0. Recalling (1.9), we do this by setting

$$b(s) := \theta(s) - \sigma s + \int_0^s \frac{c(r)a'(r)}{\theta(r)}\, dr. \tag{13.34}$$

This is justified whenever the integral on the right-hand side of (13.34) is well defined. Assuming that, for some $\nu \in (0, \delta)$, the integrals in (13.1) and (13.2) can be evaluated exactly, we then have the complete coefficients of a reaction-convection-diffusion equation for which (13.1) and (13.2) define an explicit semi-wavefront solution f with wave speed σ decreasing to 0. Moreover, if in addition $\delta < \infty$, θ satisfies the integrability condition on $[0, \delta]$, and, $\theta(\delta) = 0$, we have a reaction-convection-diffusion equation for which f is a wavefront solution from δ to 0 with wave speed σ. If $\delta = \infty$, f is an unbounded travelling-wave solution.

The method described above can be taken a step further. We can, for instance, use it to find a reaction-convection-diffusion equation with two explicit travelling-wave solutions. This works as follows. Suppose that we have a diffusion term a fulfilling Hypothesis 2.1(i), and, two different functions θ_i satisfying $\theta_i(0) = 0$ and the integrability condition on $[0, \delta)$, for some $0 < \delta \leq \ell$ and $i = 1, 2$. Suppose too that with the identical convection term b and identical reaction term c, each function θ_i corresponds to a semi-wavefront solution decreasing to 0 of equation (1.1) with wave speed σ_i. Then subtracting the integral equation (1.9) for θ_1 from that for θ_2 and differentiating, we deduce that necessarily

$$ca' = \frac{(\theta_1' - \sigma_1) - (\theta_2' - \sigma_2)}{\theta_1 - \theta_2}\theta_1\theta_2. \tag{13.35}$$

With a reaction term determined from this expression, the appropriate convection term for the sought-after partial differential equation can be found hereafter using (13.34) for θ_1 or θ_2. The resulting expression can be shown to be equivalent to

$$b' = \frac{(\theta_1' - \sigma_1)\theta_1 - (\theta_2' - \sigma_2)\theta_2}{\theta_1 - \theta_2}. \tag{13.36}$$

Application 13.11. *The equation*

$$u_t = u_{xx} + \left(3u(1 + \ln^2 u)\right)_x + \begin{cases} 0 & \textit{for } u = 0 \\ 2u(2 + \ln u)\ln^3 u & \textit{for } u > 0 \end{cases}$$

admits the following wavefront solutions from 1 *to* 0*, both with wave speed* $\sigma = -3$*:*

$$f(\xi) = \begin{cases} \exp(1/\xi) & \textit{for } \xi < 0 \\ 0 & \textit{for } \xi \geq 0 \end{cases}$$

and

$$f(\xi) = \begin{cases} \exp(1/2\xi) & \text{for } \xi < 0 \\ 0 & \text{for } \xi \geq 0. \end{cases}$$

Proof. Take $a(u) := u$, $\theta_i(s) := A_i s \ln^2 s$ for $0 < s \leq 1$ and $i = 1, 2$, where $A_1 > A_2 > 0$; and, suppose that $\sigma_1 = \sigma_2$. Then from (13.35) and (13.36), we compute $c(s) = A_1 A_2 s(2+\ln s)\ln^3 s$ and $b(s) = -\sigma_1 s + (A_1 + A_2) s \ln^2 s$ respectively. Settling for $A_1 := 2$, $A_2 := 1$ and $\sigma_1 := -3$ completes the example. □

Another way in which we could take the method described in this section a step further is the following. Starting with a function f decreasing to 0, and a number σ, we could, in principle, by assuming that f was a travelling-wave solution with wave speed σ, construct the previously-assumed function θ from it. Inspired by Applications 13.7, 13.8 and 13.11, we illustrate this with the following.

Application 13.12. *The equation*

$$u_t = u_{xx} - \begin{cases} 0 & \text{for } u = 0 \\ u(\ln u + \ln^2 u)\ln^2(-\ln u) & \text{for } 0 < u < 1 \\ 0 & \text{for } u = 1 \end{cases}$$

admits the explicit wavefront solution

$$f(\xi) = \exp\{-\exp(\exp \xi)\} \tag{13.37}$$

with wave speed $\sigma = -1$.

Proof. Differentiating (13.37) gives $-f'(\xi) = f(\xi)\{-\ln f(\xi)\}\ln\{-\ln f(\xi)\}$. Hence, if we suppose that $a(u) := u$, the function θ associated with this 'travelling wave' is

$$\theta(s) = s\,|\ln s|\ln(-\ln s). \tag{13.38}$$

Subsequently (13.31) implies $c(s) = [\sigma + 1 + \{1 + \ln s\}\ln(-\ln s)]\theta(s)$ for $0 < s < 1$. For simplicity choose $\sigma := -1$. In this case, the reaction term reduces to $c(s) = -s(\ln s + \ln^2 s)\ln^2(-\ln s)$. Voilá, we have a reaction term for a reaction-diffusion equation for which (13.37) is an explicit wavefront solution with speed $\sigma = -1$. □

Application 13.13. *The equation*

$$u_t = u_{xx} + (u\{1 + |\ln u|\ln(-\ln u)\})_x$$

admits the explicit wavefront solution (13.37) *with wave speed* $\sigma = -1$.

Proof. Assuming that $a(u) := u$, that $c(u) := 0$, and, that (13.37) is a travelling-wave solution of some equation with these diffusion and reaction terms with wave speed $\sigma := -1$; arguing as above, we once more find the 'solution' (13.38) of the integral equation. The missing convection term b in the reaction-convection-diffusion equation is subsequently immediate from (13.34). □

The approach outlined above in its totality, lends itself for further use.

Bibliographic notes

For various instances of explicit nontrivial monotonic travelling-wave solutions of nonlinear reaction-convection-diffusion equations, see [1,2,15,23,78,134,150,159, 170,177,196,211,218,225,227,231,260,261,272,303].

In the case $n = 2$, $b_0 = -1/2$ and $m = 1$, equation (13.5) is none other than the Burgers equation, and, as mentioned in Chapter 9, in this case the explicit solution in Application 13.1(c) was found by Burgers [62,63]. When $n = 2$, $b_0 = -1$ and $m = 3/2$ the equation is the foam drainage equation with a change of coordinate, and, for this particular equation the explicit solution of Application 13.1(b) was found by Gol'dfarb et al. [131] and by Verbist and Weaire [292].

Regarding Application 13.2, the explicit solution with $c_0 < 0$, $m + p = 2$, $m > 1$ and $b_0 = 0$ was exploited in the analysis of the power-law reaction-diffusion equation with a sink term by Herrero and Vázquez [161]. The explicit solutions with $c_0 > 0$, $m + p = 2$, $m > 1$ and $b_0 = 0$ were similarly used for the counterpart with a source term by de Pablo and Vázquez [228]. The derivation of the aforementioned solutions using Lie group methods and the Painlevé test is discussed in [109].

For Application 13.3, the solutions in parts (a), (b), (m) and (n), and, those in parts (p) – (s) and (u) – (x) with $m \geq 1$ and $\sigma = mc_0/nb_0$, or, with $m \geq n$ and $\sigma = mc_0/nb_0$, were found earlier by de Pablo and Sánchez [227].

According to McKean [201], to Sattinger [263], and, by implication, to Hadeler and Rothe [150], the explicit solutions embodied in Application 13.4 in the case $m = p = 1$, $b_0 = b_1 = 0$, $c_0 = 1$ and $c_1 > 0$ were obtained by Huxley. In the course of time, many of the other explicit solutions have been discovered using a variety of techniques — some systematic, some ad hoc. Often, a known solution has been rediscovered by one or more different authors. Rather than presenting an exhaustive historical survey, we summarize derivations of solutions which can be found in the literature in Tables 13.1 and 13.2. Notwithstanding, worthy of special mention are the discovery of solution (13.23) of the Fisher equation by Ablowitz and Zeppetella [2], and, the original discovery of an explicit wavefront solution displaying finite speed of propagation, for a degenerate parabolic reaction-diffusion equation, independently by Aronson [15] and Newman [218].

Application 13.8 is the reconstruction of an explicit wavefront solution of a reaction-diffusion equation known from [78]. Application 13.9 is inspired by some explicit pulse-like stationary solutions of a reaction-diffusion equation (13.30) with $a(u) = u^2$ and $c(u)$ a cubic polynomial in u found by Satsuma [260]. The application presents these solutions in a broader setting and as wavefronts.

Some of the techniques previously applied for finding travelling-wave solutions of equations of type (1.1) can quite feasibly be visualized as applications of the method outlined in Section 13.3. For instance, the technique applied by Otwinowski et al. [225] is equivalent to taking $a(u) := u$ and assuming that $\theta(s)$ is a polynomial in s of degree five; the method of Borzi et al. [52] is open to a related interpretation; the technique used by Lu et al. [196] could be viewed as assuming

Author(s)	Publication	Date	Cases			
Huxley	[201, 263]		$m = 1$	$p = 1$	$b_0 = b_1 = 0$	$c_0 = 1$, $c_1 > 0$
Hadeler & Rothe	[150]	1975	$m = 1$	$p = 1$	$b_0 = b_1 = 0$	$c_0 = 1$, $c_1 > 0$
			$m = 1$	$p = 1$	$b_0 = b_1 = 0$	$-1 < c_0 < 0$, $c_1 = 1$
Ablowitz & Zeppetella	[2, 155]	1979	$m = 1$	$p = 1/2$	$b_0 = b_1 = 0$	$c_0 = c_1 = 1$
Aronson	[15]	1980	$m = 2$	$p = 1$	$b_0 = b_1 = 0$	$c_0 = 0$, $c_1 = 1$
Newman	[218]	1980	$m > 0$	$p = 1$	$b_0 = b_1 = 0$	$c_0 = 0$, $c_1 > 0$
Parliński & Zieliński	[231]	1981	$m = 1$	$p = 1, 2, \ldots$	$b_0 = b_1 = 0$	any c_0, any c_1
Abdelkader	[1]	1982	$m = 1$	$p > 0$	$b_0 = b_1 = 0$	$c_0 = c_1 = 1$
Kawahara & Tanaka	[177]	1983	$m = 1$	$p = 1$	$b_0 = b_1 = 0$	$-1 < c_0 \leq c_1 = 1$
Newman	[219]	1983	$m > 1$	$p = m - 1$	$b_0 = b_1 = 0$	$c_0 = 0$, $c_1 > 0$
Kaliappan	[172]	1984	$m = 1$	$p > 0$	$b_0 = b_1 = 0$	$c_0 = c_1 > 0$
Satsuma	[260]	1987	$m = 2$	$p = 2$	$b_0 = b_1 = 0$	$c_0 = 0$, $c_1 = 6$
	[261]	1987	$m = 1$	$p = 1$	$b_0 = 0$, any b_1	$c_0 \leq 0$, $c_1 > \|c_0\|$
			$m = 1$	$p = 1$	$b_0 = 0$, any b_1	$c_0 \geq 0$, $c_1 < -c_0$
Otwinowski et al.	[225]	1988	$m = 1$	$p = 1$	$b_0 = b_1 = 0$	any c_0, any c_1
			$m = 1$	$p = 2$	$b_0 = b_1 = 0$	any c_0, any c_1
Wang	[71, 158, 302]	1988	$m = 1$	$p > 0$	$b_0 = b_1 = 0$	$c_0 = c_1 = 1$
Gibbs & Murray	[211]	1989	$m = 1$	$p = 2$	$b_0 = 0$, $b_1 < 0$	$c_0 = 0$, $c_1 > 0$
Cariello & Tabor	[65]	1989	$m = 1$	$p = 1$	$b_0 = b_1 = 0$	$c_0 = c_1 = 1$
Murray	[211]	1989	$m = 1$	$p > 0$	$b_0 = b_1 = 0$	$c_0 = c_1 = 1$
			$m = 1$	$p > 0$	$b_0 = b_1 = 0$	$c_0 = 0$, $c_1 = 1$
			$m = 1$	$p = 1$	$b_0 = b_1 = 0$	$c_0 < 0$, $c_1 > \|c_0\|$
Abur-Robb	[3, 4]	1990	$m = 1$	$p = 1$	$b_0 = b_1 = 0$	$-1 < c_0 < 0$, $c_1 = 1$
Hereman	[154]	1990	$m = 1$	$p = 1$	$b_0 = b_1 = 0$	$c_0 < 0$, $c_1 > 0$
Wang et al.	[303]	1990	$m = 1$	$p > 0$	$b_0 = 0$, any b_1	$c_0 < 0$, $c_1 > \|c_0\|$

Table 13.1: Early discoveries of explicit wavefront solutions listed in Application 13.4.

Author(s)	Publication	Date	Cases			
Grindrod	[134]	1991	$m = 1$	$p = 1$	$b_0 = b_1 = 0$	any c_0, $c_1 = 1$
			$m = 2$	$p = 1$	$b_0 = b_1 = 0$	$c_0 = 0$, $c_1 = 1$
Dixon et al.	[83]	1991	$m = 1$	$p > 0$	$b_0 = b_1 = 0$	any c_0, any c_1
Choudhury	[69]	1991	$m = 1$	$p > 0$	$b_0 = b_1 = 0$	$c_0 = 0$, $c_1 > 0$
			$m = 1$	$p > 0$	$b_0 = b_1 = 0$	$c_0 = c_1 > 0$
Chen & Guo	[67]	1992	$m = 1$	$p = 1$	$b_0 = b_1 = 0$	$-1 < c_0 \leq c_1 = 1$
			$m = 2$	$p = 1$	$b_0 = b_1 = 0$	$c_0 = 0$, $c_1 = 1$
Herrera et al.	[159]	1992	$m = 2$	$p = 1$	$b_0 = b_1 = 0$	$c_0 = 0$, $c_1 = 1$
			$m = 2$	$p = 2$	$b_0 = b_1 = 0$	$c_0 = 0$, $c_1 = 1$
			$m = 3$	$p = 2$	$b_0 = b_1 = 0$	$c_0 = 0$, $c_1 = 1$
			$m = 3$	$p = 4$	$b_0 = b_1 = 0$	$c_0 = 0$, $c_1 = 1$
			$m = 1/2$	$p = -1/2$	$b_0 = b_1 = 0$	$c_0 = 0$, $c_1 = 1$
Malfliet	[198]	1992	$m = 2$	$p = 1$	$b_0 = 0$, $b_1 < 0$	$c_0 = 0$, $c_1 = 1$
Lu et al.	[196]	1993	$m = 1$	$p > -1$	$b_0 = b_1 = 0$	$c_0 = c_1 > 0$
			$m = 1$	$p > -1$	$b_0 = b_1 = 0$	$c_0 = 0$, $c_1 > 0$
			$m > 0$	$p = m - 1$	$b_0 = b_1 = 0$	$c_0 = 0$, $c_1 > 0$
			$m = 1$	$p > 0$	$b_0 = 0$, any b_1	$c_0 > c_1 = 0$
			$m = 1$	$p > 0$	$b_0 = 0$, any b_1	$c_0 < 0$, $c_1 > \lvert c_0 \rvert$
Abur-Robb	[5]	1994	$m = 1$	$p = 1$	$b_0 = b_1 = 0$	$c_0 < 0$, $c_1 > 0$
Wang et al.	[301]	1994	$m = 1$	$p > 0$	$b_0 = b_1 = 0$	any c_0, $c_1 > 0$
Yang et al.	[311]	1994	$m = 1$	$p = 1$	$b_0 = b_1 = 0$	$c_0 > c_1 > 0$
Danilov et al.	[78]	1995	$m > 1$	$p = m - 1$	$b_0 = b_1 = 0$	$c_0 = c_1 > 0$
Gudkov	[137]	1997	$m = 1$	$p = 1$	$b_0 = 0$, any b_1	any c_0, any c_1
Kessler et al.	[183]	1998	$m = 1$	$p = 1$	$b_0 = b_1 = 0$	$c_0 < 0$, $c_1 > 2 \lvert c_0 \rvert$
Rosenau	[247]	2002	$m > 1$	$p = m - 1$	$b_0 = 0$, any b_1	$c_0 = 0$, $c_1 = 1$

Table 13.2: Late discoveries of explicit wavefront solutions listed in Application 13.4.

that $a(u) := u^m$ and $\theta(s) := s(\theta_0 + \theta_1 s^p)$ for a selected set of parameters m, p, θ_0 and θ_1, and, essentially the method used by Benguria and Depassier [34] for finding wavefront solutions from 1 to 0 for the equation with $a(u) := u$ and $c(u)$ a polynomial of degree five assumes $\theta(s) := s(1-s)(\theta_0 + \theta_1 s)$ for some numbers θ_0 and θ_1. The approach of Yang [310] boils done to taking $a(u) := u$, $b(u) := 0$ and $\theta(s) := \theta_0 s + \theta_s^p$ for some $\theta_0 > 0$, $\theta_1 > 0$ and $p > 0$, $p \neq 1$. The method advocated by Ebert and van Saarloos [94] is without doubt the same as that used in Section 13.3.

The approach advocated in Section 13.3 of deriving a set of coefficients corresponding to a given 'solution' was employed earlier by Philip [236–238] in his discovery of explicit self-similar solutions of the nonlinear diffusion equation, $u_t = (a(u))_{xx}$, via the Boltzmann transformation.

Bibliography

[1] M.A. Abdelkader, *Travelling wave solutions for a generalized Fisher equation*, J. Math. Anal. Appl. **85** (1982), 287–290.

[2] M.J. Ablowitz & A. Zeppetella, *Explicit solutions of Fisher's equation for a special wave speed*, Bull. Math. Biol. **41** (1979), 835–840.

[3] M.F.K. Abur-Robb, *Explicit solutions of Fisher's equation with three zeros*, Internat. J. Math. Math. Sci. **13** (1990), 617–620.

[4] M.F.K. Abur-Robb, *Self-solutions of Fisher's equation with three zeros by operator decomposition*, Dirasat Ser. B **19** (1992), 59–62.

[5] M.F.K. Abur-Robb, *Self-similar transformations for Fisher-type equations*, Dirasat Ser. B **21** (1994), 292–298.

[6] J. Aguirre & M. Escobedo, *A Cauchy problem for $u_t - \Delta u = u^p$ with $0 < p < 1$. Asymptotic behaviour of solutions*, Ann. Fac. Sci. Toulouse Math. (5) **8** (1978), 175–203.

[7] A.P. Aldushin, S.I. Khudyaev & Y.B. Zel'dovich, *Flame propagation in the reacting gaseous mixture*, Arch. Combustionis **1** (1981), 9–21.

[8] L. Alvarez & J.I. Diaz, *Sufficient and necessary initial mass conditions for the existence of a waiting time in nonlinear-convection processes*, J. Math. Anal. Appl. **155** (1991), 378–392.

[9] L. Alvarez, J.I. Diaz & R. Kersner, *On the initial growth of the interfaces in nonlinear diffusion-convection processes*, Nonlinear Diffusion Equations and Their Equilibrium States I (edited by W.-M. Ni, L.A. Peletier & J. Serrin), Springer-Verlag, New York, 1988, pp. 1–20.

[10] W.F. Ames, *Nonlinear Partial Differential Equations in Engineering*, Academic Press, New York, 1965.

[11] W.F. Ames, *Nonlinear Partial Differential Equations in Engineering Volume II*, Academic Press, New York, 1972.

[12] V.K. Andreev, O.V. Kaptsov, V.V. Pukhnachov & A.A. Rodionov, *Applications of Group-Theoretical Methods in Hydrodynamics*, Kluwer Academic Publishers, Dordrecht, The Netherlands, 1998.

[13] R. Aris, *The Mathematical Theory of Diffusion and Reaction in Permeable Catalysts, Volume I, The Theory of the Steady State*, Clarendon Press, Oxford, 1975.

[14] R. Aris, *The Mathematical Theory of Diffusion and Reaction in Permeable Catalysts, Volume II, Questions of Uniqueness, Stability, and Transient Behaviour*, Clarendon Press, Oxford, 1975.

[15] D.G. Aronson, *Density-dependent interaction-diffusion systems*, Dynamics and Modelling of Reactive Systems (edited by W.E. Stewart, W.H. Ray & C.C. Conley), Academic Press, New York, 1980, pp. 161–176.

[16] D.G. Aronson, *The porous medium equation*, Nonlinear Diffusion Problems (edited by A. Fasano & M. Primicerio), Lecture Notes in Mathematics **1224**, Springer-Verlag, Berlin, 1986, pp. 1–46.

[17] D.G. Aronson, *Regularity of flows in porous media: A survey*, Nonlinear Diffusion Equations and Their Equilibrium States I (edited by W.-M. Ni, L.A. Peletier & J. Serrin), Springer-Verlag, New York, 1988, pp. 35–49.

[18] D.G. Aronson & J.L. Vazquez, *Calculation of anomalous exponents in nonlinear diffusion*, Phys. Rev. Lett. **72** (1994), 348–351.

[19] D.G. Aronson & J.L. Vazquez, *Anomalous exponents in nonlinear diffusion*, J. Nonlinear Sci. **5** (1995), 29–56.

[20] D.G. Aronson & H.F. Weinberger, *Nonlinear diffusion in population genetics, combustion, and nerve pulse propagation*, Partial Differential Equations and Related Topics (edited by J.A. Goldstein), Lecture Notes in Mathematics **446**, Springer-Verlag, Berlin, 1975, pp. 5–49.

[21] D.G. Aronson & H.F. Weinberger, *Multidimensional nonlinear diffusion arising in population genetics*, Adv. Math. **30** (1978), 33–76.

[22] C. Atkinson, G.E.H. Reuter & C.J. Ridler-Rowe, *Traveling wave solutions for some nonlinear diffusion equations*, SIAM J. Math. Anal. **12** (1981), 880–892.

[23] V. Barelko, I.I. Kurochka, A.G. Merzhanov & K.G. Shkadinskii, *Investigation of travelling waves on catalytic wires*, Chem. Engrg. Sci. **33** (1978), 805–811.

[24] G.I. Barenblatt, *Self-similarity: Similarity and intermediate asymptotic form*, Radiophys. Quantum Electron. **19** (1976), 643–664. Translation of: Izv. Vyssh. Uchebn. Zaved. Radiofiz. **19** (1976), 902–931.

[25] G.I. Barenblatt, *The diffusion equation*, Selected Works of A.N. Kolmogorov Volume I Mathematics and Mechanics (edited by V.M. Tikhomirov), Kluwer Academic Publishers, Dordrecht, The Netherlands, 1991, pp. 481–487.

[26] G.I. Barenblatt & M.I. Vishik, *On finite velocity of propagation in problems of non-stationary filtration of a liquid or gas* (in Russian), Prikl. Mat. Mekh. **20** (1956), 411–417.

[27] G.I. Barenblatt & Ya.B. Zel'dovich, *Intermediate asymptotics in mathematical physics*, Russian Math. Surveys **26** (1971), 45–61. Translation of: Uspekhi Mat. Nauk **26** (1971), 115–129.

[28] G.I. Barenblatt & Ya.B. Zel'dovich, *Self-similar solutions as intermediate asymptotics*, Annual Rev. Fluid Mech. **4** (1972), 285–312.

[29] P.W. Bates, *Travelling waves in radially symmetric parabolic systems*, Proc. Roy. Soc. Edinburgh Sect. A **99** (1985), 269–275.

[30] J. Bear, *Dynamics of Fluids in Porous Media*, American Elsevier Publishing Company, New York, 1972.

[31] J. Bebernes & D. Eberly, *Mathematical Problems from Combustion Theory*, Springer-Verlag, New York, 1989.

[32] R.D. Benguria, J. Cisternas & M.C. Depassier, *Variational calculations for thermal combustion waves*, Phys. Rev. E **52** (1995), 4410–4413.

[33] R.D. Benguria & M.C. Depassier, *Validity of the linear speed selection mechanism for fronts of the nonlinear diffusion equation*, Phys. Rev. Lett. **73** (1994), 2272–2274.

[34] R.D. Benguria & M.C. Depassier, *Exact fronts for the nonlinear diffusion equation with quintic nonlinearities*, Phys. Rev. E **50** (1994), 3701–3704.

[35] R.D. Benguria & M.C. Depassier, *Variational principle for the asymptotic speed of fronts of the density-dependent diffusion-reaction equation*, Phys. Rev. E **52** (1995), 3285–3287.

[36] R.D. Benguria & M.C. Depassier, *Variational characterization of the speed of propagation of fronts for the nonlinear diffusion equation*, Comm. Math. Phys. **175** (1996), 221–227.

[37] R.D. Benguria & M.C. Depassier, *Speed of fronts of the reaction-diffusion equation*, Phys. Rev. Lett. **77** (1996), 1171–1173.

[38] R.D. Benguria & M.C. Depassier, *Monotonic fronts of the reaction-diffusion equation in one space dimension*, Instabilities and Nonequilibrium Structures V (edited by E. Tirapegui & W. Zeller, Kluwer Academic Publishers, Dordrecht, The Netherlands, 1996, pp. 105–110.

[39] R.D. Benguria & M.C. Depassier, *Speed of fronts of generalized reaction-diffusion equations*, Phys. Rev. E **57** (1998), 6493–6496.

[40] E. Ben-Jacob, H. Brand, G. Dee, L. Kramer & J.S. Langer, *Pattern propagation in nonlinear dissipative systems*, Phys. D **14** (1985), 348–364.

[41] E. Ben-Jacob, I. Cohen & H. Levine, *Cooperative self-organization of microorganisms*, Adv. Phys. **49** (2000), 395–554.

[42] H. Berestycki, B. Nicolaenko & B. Scheurer, *Traveling wave solutions to reaction diffusion systems modeling combustion*, Contemp. Math. **17** (1983), 189–208.

[43] H. Berestycki, B. Nicolaenko & B. Scheurer, *Traveling wave solutions to combustion models and their singular limits*, SIAM J. Math. Anal. **16** (1985), 1207–1242.

[44] H. Berestycki & L. Nirenberg, *Travelling front solutions of semilinear equations in n dimensions*, Frontiers in Pure and Applied Mathematics (edited by R. Dautray), North-Holland, Amsterdam, 1991, pp. 31–41.

[45] H. Berestycki & L. Nirenberg, *Travelling fronts in cylinders*, Ann. Inst. H. Poincaré Anal. Non Linéaire **9** (1992), 497–572.

[46] J.G. Berryman, *Evolution of stable profile for a class of nonlinear diffusion equations with fixed boundaries*, J. Math. Phys. **18** (1977), 2108–2115.

[47] J. Billingham & D.J. Needham, *A note on the properties of a family of travelling-wave solutions arising in cubic autocatalysis*, Dynamics and Stability of Systems **6** (1991), 33–49.

[48] Z. Biro, *Attractors in a density-dependent diffusion-reaction model*, Nonlinear Anal. **29** (1997), 485–499.

[49] Z. Biró, *Stability of travelling waves for degenerate reaction-diffusion equations of KPP-type*, Adv. Nonlinear Stud. **2** (2002), 357–371.

[50] G.H. Bolt, *Movement of solutes in soil: Principles of adsorption/exchange chromatography*, Soil Chemistry B. Physico-Chemical Models (edited by G.H. Bolt), Elsevier Scientific Publishing Company, Amsterdam, 1979, pp. 285–348.

[51] M.R. Booty, R. Haberman & A.A. Minzoni, *The accommodation of traveling waves of Fisher's type to the dynamics of the leading tail*, SIAM J. Appl. Math. **53** (1993), 1009–1025.

[52] C. Borzi, H.L. Frisch, R. Gianotti & J.K. Percus, *Inverse solution for some travelling-wave reaction-diffusion problems*, J. Phys. A **23** (1990), 4823–4830.

[53] M. Bramson, *Convergence of solutions of the Kolmogorov equation to travelling waves*, Mem. Amer. Math. Soc. **44** (1983), Number 285.

[54] M. Bramson, *Location of the travelling wave for the Kolmogorov equation*, Probab. Theory Related Fields **73** (1986), 481–515.

[55] M.D. Bramson, *Maximal displacement of branching Brownian motion*, Comm. Pure Appl. Math. **31** (1978), 531–581.

[56] N.F. Britton, *Reaction-Diffusion Equations and Their Applications to Biology*, Academic Press, London, 1986.

[57] J. Buckmaster, *Viscous sheets advancing over dry beds*, J. Fluid Mech. **81** (1977), 735–756.

[58] J.D. Buckmaster & G.S.S. Ludford, *Theory of Laminar Flames*, Cambridge University Press, Cambridge, U.K., 1982.

[59] J.D. Buckmaster & G.S.S. Ludford, *Lectures on Mathematical Combustion*, Society for Industrial and Applied Mathematics, Philadelphia, Penn., 1983.

[60] J.M. Burgers, *Mathematical examples illustrating relations occurring in the theory of turbulent fluid motion*, Verh. Nederl. Akad. Wetensch. Afd. Natuurk. Sect. 1 **17** (1939), 1–53.

[61] J.M. Burgers, *Application of a model system to illustrate some points of the statistical theory of free turbulence*, Proc. Kon. Nederl. Akad. Wetensch. **43** (1940), 2–12.

[62] J.M. Burgers, *A mathematical model illustrating the theory of turbulence*, Advances in Applied Mechanics **1** (1948), 171–199.

[63] J.M. Burgers, *The Nonlinear Diffusion Equation*, D. Reidel Publishing Company, Dordrecht, Holland, 1975.

[64] J. Canosa, *On a nonlinear diffusion equation describing population growth*, IBM J. Res. Develop. **17** (1973), 307–313.

[65] F. Cariello & M. Tabor, *Painlevé expansions for nonintegrable evolution equations*, Phys. D **39** (1989), 77–94.

[66] F.F. Chen, *Introduction to Plasma Physics*, Plenum Press, New York, 1974.

[67] Chen Z. & Guo B., *Analytic solutions of the Nagumo equation*, IMA J. Appl. Math. **48** (1992), 107–115.

[68] E.C. Childs, *An Introduction to the Physical Basis of Soil Water Phenomena*, John Wiley & Sons, London, 1969.

[69] S.R. Choudhury, *Painlevé analysis and special solutions of two families of reaction-diffusion equations*, Phys. Lett. A **159** (1991), 311–317.

[70] J. Cisternas & M.C. Depassier, *Counterexample to a conjecture of Goriely for the speed of fronts of the reaction-diffusion equation*, Phys. Rev. E **55** (1997), 3701–3704.

[71] M.W. Coffey, *Phase space analysis for the direct algebraic method for nonlinear evolution and wave equations*, SIAM J. Appl. Math. **52** (1992), 929–945.

[72] E. Comparini, R. Ricci & J.L. Vazquez, *Asymptotic behavior of the solutions of a nonlinear Fokker-Planck equation with Dirichlet boundary conditions*, J. Math. Anal. Appl. **175** (1993), 606–631.

[73] T.B. Constable & N.T.S. Evans, *A method for measuring the spatial distribution of tissue oxygen removal rates*, Respiration Physiology **25** (1975), 175–190.

[74] J. Crank, *The Mathematics of Diffusion*, Clarendon Press, Oxford, 1975.

[75] J. Crank, *Free and Moving Boundary Problems*, Clarendon Press, Oxford, 1984.

[76] J. Crank & R.S. Gupta, *A moving boundary problem arising from the diffusion of oxygen in absorbing tissue*, J. Inst. Math. Appl. **10** (1972), 19–33.

[77] M.C. Cross & P.C. Hohenberg, *Pattern formation outside of equilibrium* Rev. Modern Phys. **65** (1993), 851–1112.

[78] V.G. Danilov, V.P. Maslov & K.A. Volosov, *Mathematical Modelling of Heat and Mass Transfer Processes*, Kluwer Academic Publishers, Dordrecht, The Netherlands, 1995.

[79] G. Dee & J.S. Langer, *Propagating pattern selection*, Phys. Rev. Lett. **50** (1983), 383–386.

[80] J.I. Diaz & R. Kersner, *Non existence d'une des frontières libres dans une équation dégénérée en théorie de la filtration*, C. R. Acad. Sci. Paris Sér. I Math. **296** (1983), 505–508.

[81] J.I. Diaz & R. Kersner, *On the behaviour and cases of nonexistence of the free boundary in a semibounded porous medium*, J. Math. Anal. Appl. **132** (1988), 281–289.

[82] O. Diekmann & N.M. Temme (editors), *Nonlinear Diffusion Problems*, Mathematisch Centrum, Amsterdam, 1976.

[83] J.M. Dixon, J.A. Tuszyński & M. Otwinowski, *Special analytical solutions of the damped-anharmonic-oscillator equation*, Phys. Rev. A **44** (1991), 3484–3491.

[84] M. Dong & I. Chatzis, *The imbibition and flow of a wetting liquid along the corners of a square capillary tube*, J. Colloid Interface Sci. **172** (1995), 278–288.

[85] C.J. van Duijn & J.M. de Graaf, *Large time behaviour of solutions of the porous medium equation with convection*, J. Differential Equations **84** (1990), 183–203.

[86] C.J. van Duijn & P. Knabner, *Solute transport through porous media with slow adsorption*, Free Boundary Problems: Theory and Applications Volume I (edited by K.H. Hoffmann & J. Sprekels), Longman Scientific & Technical, Harlow, England, 1990, pp. 375–388.

[87] C.J. van Duijn & P. Knabner, *Solute transport in porous media with equilibrium and non-equilibrium multiple-site absorption: Travelling waves*, J. Reine Angew. Math. **415** (1991), 1–49.

[88] C.J. van Duijn & P. Knabner, *Travelling waves in the transport of reactive solutes through porous media: Adsorption and binary ion exchange — Part 1*, Transport in Porous Media **8** (1992), 167–194.

[89] C.J. van Duijn & P. Knabner, *Travelling waves in the transport of reactive solutes through porous media: Adsorption and binary ion exchange — Part 2*, Transport in Porous Media **8** (1992), 199–225.

[90] C.J. van Duijn & P. Knabner, *Travelling wave behaviour of crystal dissolution in porous media flow*, European J. Appl. Math. **8** (1997), 49–72.

[91] C.J. van Duijn, P. Knabner & S.E.A.T.M. van der Zee, *Travelling waves during the transport of reactive solute in porous media: Combination of Langmuir and Freundlich isotherms*, Adv. Water Resources **16** (1993), 97–105.

[92] G. Duvaut, *Résolution d'un problème de Stefan (Fusion d'un bloc de glace á zèro degré)*, C. R. Acad. Sci. Paris Sér. A **276** (1973), 1461–1463.

[93] C.J. van Duyn & J.M. de Graaf, *Limiting profiles in contaminant transport through porous media*, SIAM J. Math. Anal. **18** (1987), 728–743.

[94] U. Ebert & W. van Saarloos, *Front propagation into unstable states: Universal algebraic convergence towards uniformly translating pulled fronts*, Phys. D **146** (2000), 1–99.

[95] J.-P. Eckmann & C.E. Wayne, *The non-linear stability of front solutions for parabolic partial differential equations*, Comm. Math. Phys. **161** (1994), 323–334.

[96] C.M. Elliott & J.R. Ockendon, *Weak and Variational Methods for Moving Boundary Problems*, Pitman Advanced Publishing Program, Boston, 1982.

[97] H. Engler, *Relations between travelling wave solutions of quasilinear parabolic equations*. Proc. Amer. Math. Soc. **93** (1985), 297–302.

[98] P.C. Fife, *Asymptotic states for equations of reaction and diffusion*, Bull. Amer. Math. Soc. (N.S.) **84** (1978), 693–726.

[99] P.C. Fife, *Long time behavior of solutions of bistable nonlinear diffusion equations*, Arch. Rational Mech. Anal. **70** (1979), 31–46.

[100] P.C. Fife, *The bistable nonlinear diffusion equation: Basic theory and some applcations*, Applied Nonlinear Analysis (edited by V. Lakshmikantham), Academic Press, New York, 1979, pp. 143–160.

[101] P.C. Fife, *Mathematical Aspects of Reacting and Diffusing Systems*, Lecture Notes in Biomathematics **28**, Springer-Verlag, Berlin, 1979.

[102] P.C. Fife & J.B. McLeod, *The approach of solutions of nonlinear diffusion equations to travelling wave solutions*, Bull. Amer. Math. Soc. (N.S.) **81** (1975), 1076–1078.

[103] P.C. Fife & J.B. McLeod, *The approach of solutions of nonlinear diffusion equations to travelling front solutions*, Arch. Rational Mech. Anal. **65** (1977), 335–361.

[104] P.C. Fife & J.B. McLeod, *A phase plane discussion of convergence to travelling fronts for nonlinear diffusion*, Arch. Rational Mech. Anal. **75** (1981), 281–314.

[105] R.A. Fisher, *The wave of advance of advantageous genes*, Annals of Eugenics **7** (1937), 355–369.

[106] D.A. Frank-Kamenetskii, *Diffusion and Heat Transfer in Chemical Kinetics*, Plenum Press, New York, 1969.

[107] M. Freidlin, *Functional Integration and Partial Differential Equations*, Princeton University Press, Princeton, N.J., 1985.

[108] M.I. Freidlin, *Geometric optics approach to reaction-diffusion equations*, SIAM J. Appl. Math. **46** (1986), 222–232.

[109] H.-D. Frey, W.G. Glöckle & T.F. Nonnenmacher, *Symmetries and integrability of generalized diffusion reaction equations*, J. Phys. A **26** (1993), 665–679.

[110] V.A. Galaktionov, *Some properties of travelling waves in a medium with nonlinear thermal conductivity and a source of heat*, U.S.S.R. Comput. Math. and Math. Phys. **21** (1981), 167–176. Translation of: Zh. Vychisl. Mat. i Mat. Fiz. **21** (1981), 980–989.

[111] V.A. Galaktionov, *Geometric theory of one-dimensional nonlinear parabolic equations I. Singular interfaces*, Adv. Differential Equations **7** (2002), 513–580.

[112] V.A. Galaktionov, V.A. Dorodnitsyn, G.G. Elenin, S.P. Kurdyumov & A.A. Samarskii, *A quasilinear heat equation with a source: Peaking, localization, symmetry exact solutions, asymptotics, structures*, J. Soviet Math. **41** (1988), 1222–1292. Translation of: Itogi Nauki i Tekniki **28** (1986), 95–206.

[113] V.A. Galaktionov & R. Kersner, *Discontinuous limit semigroups for the singular Zhang equation and its hydrodynamic version*, Adv. Differential Equations **6** (2001), 1493–1516.

[114] V.A. Galaktionov & J.L. Vazquez, *Necessary and sufficient conditions for complete blow-up and extinction for one-dimensional quasilinear heat equations*, Arch. Rational Mech. Anal. **129** (1995), 225–244.

[115] J. Gärtner, *Location of wave fronts for the multi-dimensional K-P-P equation and Brownian first exit densities*, Math. Nachr. **105** (1982), 317–351.

[116] B.H. Gilding, *Properties of solutions of an equation in the theory of infiltration*, Arch. Rational Mech. Anal. **65** (1977), 203–225.

[117] B.H. Gilding, *A nonlinear degenerate parabolic equation*, Ann. Scuola Norm. Sup. Pisa Cl. Sci. (4) **4** (1977), 393–432.

[118] B.H. Gilding, *Similarity solutions of the porous media equation*, J. Hydrology **56** (1982), 251–263.

[119] B.H. Gilding, *The occurrence of interfaces in nonlinear diffusion-advection processes*, Arch. Rational Mech. Anal. **100** (1988), 243–263.

[120] B.H. Gilding, *Improved theory for a nonlinear degenerate parabolic equation*, Ann. Scuola Norm. Sup. Pisa C1. Sci (4) **16** (1989), 165–224; ibid **17** (1990), 479.

[121] B.H. Gilding, *Localization of solutions of a nonlinear Fokker-Planck equation with Dirichlet boundary conditions*, Nonlinear Anal. **13** (1989), 1215–1240.

[122] B.H. Gilding, *Qualitative mathematical analysis of the Richards equation*, Transport in Porous Media **5** (1991), 651–666.

[123] B.H. Gilding, *A singular nonlinear Volterra integral equation*, J. Integral Equations Appl. **5** (1993), 465–502.

[124] B.H. Gilding, *The correspondence between travelling-wave solutions of a nonlinear reaction-convection-diffusion equation and an integral equation*, Differential Integral Equations **9** (1996), 919–947.

[125] B.H. Gilding & R. Kersner, *Instantaneous shrinking in nonlinear diffusion-convection*, Proc. Amer. Math. Soc. **109** (1990), 385–394.

[126] B.H. Gilding & R. Kersner, *Diffusion-convection-réaction, frontières libres et une équation intégrable*, C. R. Acad. Sci. Paris Sér. I Math. **313** (1991), 743–746.

[127] B.H. Gilding & R. Kersner, *The characterization of reaction-convection-diffusion processes by travelling waves*, J. Differential Equations **124** (1996), 27–79.

[128] B.H. Gilding & R. Kersner, *Instantaneous extinction, step discontinuities and blow-up*, Nonlinearity **16** (2003), 843–854.

[129] B.H. Gilding, R. Natalini & A. Tesei, *How parabolic free boundaries approximate hyperbolic fronts*, Trans. Amer. Math. Soc. **352** (2000), 1797–1824.

[130] B.H. Gilding & L.A. Peletier, *The Cauchy problem for an equation in the theory of infiltration*, Arch. Rational Mech. Anal. **61** (1976), 127–140.

[131] I.I. Gol'dfarb, K.B. Kann & I.R. Shreiber, *Liquid flow in foams*, Fluid Dynamics **23** (1988), 244–249. Translation of: Izv. Akad. Nauk SSSR Mekh. Zhidk. Gaza (1988), 103–108.

[132] M.O. Gomez, C.-M. Chang & V.S. Manoranjan, *Exact solution for contaminant transport with nonlinear sorption*, Appl. Math. Lett. **9** (1996), 83–87.

[133] A. Goriely, *Simple solution to the nonlinear front problem*, Phys. Rev. Lett. **75** (1995), 2047–2050.

[134] P. Grindrod, *Patterns and Waves The Theory and Applications of Reaction-Diffusion Equations*, Clarendon Press, Oxford, 1991.

[135] P. Grindrod & B.D. Sleeman, *Weak travelling fronts for population models with density dependent dispersion*, Math. Meth. Appl. Sci. **9** (1987), 576–586.

[136] G. Gripenberg, S.-O. Londen & O. Staffans, *Volterra Integral and Functional Equations*, Cambridge University Press, Cambridge, 1990.

[137] V.V. Gudkov, *A family of exact travelling wave solutions to nonlinear evolution and wave equations*, J. Math. Phys. **38** (1997), 4794–4803.

[138] W.S.C. Gurney & R.M. Nisbet, *The regulation of inhomogeneous populations*, J. Theoret. Biol. **52** (1975), 441–457.

[139] M.E. Gurtin *Thermomechanics of Evolving Phase Boundaries in the Plane*, Clarendon Press, Oxford, 1993.

[140] M.E. Gurtin & R.C. MacCamy, *On the diffusion of biological populations*, Math. Biosci. **33** (1977), 35–49.

[141] K.P. Hadeler, *Travelling population fronts*, Population Genetics and Ecology (edited by S. Karlin & E. Nevo), Academic Press, New York, 1976, pp. 585–592.

[142] K.P. Hadeler, *Nonlinear diffusion equations in biology*, Ordinary and Partial Differential Equations (edited by W.N. Everitt & B.D. Sleeman), Lecture Notes in Mathematics **564**, Springer-Verlag, Berlin, 1976, pp. 163–206.

[143] K.P. Hadeler, *Travelling fronts and free boundary value problems*, Numerical Treatment of Free Boundary Value Problems (edited by J. Albrecht, L. Collatz & K.-H. Hoffmann) Birkhauser Verlag, Basel, 1982, pp. 90–107.

[144] K.P. Hadeler, *Nonlinear differential equations from biology*, Nonlinear Analysis, Function Spaces and Applications Volume 2 (edited by O. John & A. Kufner), Teubner Verlagsgesellschaft, Leipzig, 1982, pp. 93–125.

[145] K.P. Hadeler, *Free boundary problems in biological models*, Free Boundary Problems: Theory and Applications Volume II (edited by A. Fasano & M. Primicerio) Pitman Advanced Publishing Program, Boston, 1983, pp. 664–671.

[146] K.P. Hadeler, *Traveling fronts in parabolic and hyperbolic equations*, Dynamical Systems (edited by A.B. Kurzhanski & K. Sigmund), Lecture Notes in Economics and Mathematical Systems **287**, Springer-Verlag, Berlin, 1987, pp. 154–164.

[147] K.P. Hadeler, *Hyperbolic travelling fronts*, Proc. Edinburgh Math. Soc. **31** (1988), 89–97.

[148] K.P. Hadeler, *Travelling fronts for correlated random walks*, Canadian Appl. Math. Quart. **2** (1994), 27–43.

[149] K.P. Hadeler & E.T. Poulsen, *The asymptotic displacement of a travelling front*, Methoden Verfahren Math. Phys. **17** (1977), 141–151.

[150] K.P. Hadeler & F. Rothe, *Travelling fronts in nonlinear diffusion equations*, J. Math. Biol. **2** (1975), 251–263.

[151] P.S. Hagan, *Traveling wave and multiple traveling wave solutions of parabolic equations*, SIAM J. Math. Anal. **13** (1982), 717–738.

[152] H. Haken, *Synergetics*, Springer-Verlag, Berlin, 1983.

[153] C.K. Hayes, *A new traveling-wave solution of Fisher's equation with density-dependent diffusivity*, J. Math. Biol. **29** (1991), 531–537.

[154] W. Hereman, *Application of a MACSYMA program for the Painlevé test to the Fitzhugh-Nagumo equation*, Partially Integrable Evolution Equations in Physics (edited by R. Conte & N. Boccara), Kluwer Academic Publishers, Dordrecht, The Netherlands, 1990, pp. 585–586.

[155] W. Hereman & S. Angenent, *The Painlevé test for nonlinear ordinary and partial differential equations*, MACSYMA Newsletter **6** (1989), 11–18.

[156] W. Hereman, P.P. Banerjee, A. Korpel, G. Assanto, A. van Immerzeele & A. Meerpoel, *Exact solitary wave solutions of non-linear evolution and wave equations using a direct algebraic method*, J. Phys. A **19** (1986), 607–628.

[157] W. Hereman, A. Korpel & P.P. Banerjee, *A general physical approach to solitary wave construction from linear solutions*, Wave Motion **7** (1985), 283–289.

[158] W. Hereman & M. Takaoka, *Solitary wave solutions of nonlinear evolution and wave equations using a direct method and MACSYMA*, J. Phys. A **23** (1990), 4805–4822.

[159] J.J.E. Herrera, A. Minzoni & R. Ondarza, *Reaction-diffusion equations in one dimension: Particular solutions and relaxation*, Phys. D **57** (1992), 249–266.

[160] M.A. Herrero & J.L. Vázquez, *The one-dimensional nonlinear heat equation with absorption: Regularity of solutions and interfaces*, SIAM J. Math. Anal. **18** (1987), 149–167.

[161] M.A. Herrero & J.L. Vazquez, *Thermal waves in absorbing media*, J. Differential Equations **74** (1988), 218–233.

[162] F. Hoppensteadt, *Mathematical Theories of Populations: Demographics, Genetics and Epidemics*, Society for Industrial and Applied Mathematics, Philadelphia, 1975.

[163] Y. Hosono, *Traveling wave solutions for some density dependent diffusion equations*, Japan J. Appl. Math. **3** (1986), 163–196.

[164] A.M. Il'in & O.A. Oleinik, *Behaviour of solutions of the Cauchy problem for certain quasilinear equations for unbounded increase of time* (in Russian), Dokl. Akad. Nauk SSSR **120** (1958), 25–28.

[165] A.M. Il'in & O.A. Oleinik, *Asymptotic behaviour of solutions of the Cauchy problem for some quasilinear equations for large values of time* (in Russian), Mat. Sb. **51** (1960), 191–216.

[166] W.E. Johnson, *On a first-order boundary value problem from laminar flame theory*, Arch. Rational Mech. Anal. **13** (1963), 46–54.

[167] W.E. Johnson & W. Nachbar, *Laminar flame theory and the steady, linear burning of a monopropellant*, Arch. Rational Mech. Anal. **12** (1963), 58–92.

[168] C.K.R.T. Jones, *Spherically symmetric solutions of a reaction-diffusion equation*, J. Differential Equations **49** (1983), 142–169.

[169] C.K.R.T. Jones, R. Gardner & T. Kapitula, *Stability of travelling waves for non-convex scalar viscous conservation laws*, Comm. Pure Appl. Math. **46** (1993), 505–526.

[170] D.S. Jones & B.D. Sleeman, *Differential Equations and Mathematical Biology*, George Allen and Unwin, London, 1983.

[171] A.S. Kalashnikov, *Some problems of the qualitative theory of non-linear degenerate second-order parabolic equations*, Russian Math. Surveys **42** (1987), 169–222. Translation of: Uspekhi Mat. Nauk **42** (1987), 135–176.

[172] P. Kaliappan, *An exact solution for travelling waves of* $u_t = Du_{xx} + u - u^k$, Phys. D **11** (1984), 368–374.

[173] Y. Kametaka, *On the nonlinear diffusion equation of Kolmogorov-Petrovskii-Piskunov type*, Osaka J. Math. **13** (1976), 11–66.

[174] Ya.I. Kanel', *The behavior of solutions of the Cauchy problem when the time tends to infinity, in the case of quasilinear equations arising in the theory of combustion*, Soviet Math. Dokl. **1** (1960), 533–536. Translation of: Dokl. Akad. Nauk SSSR **132** (1960), 268–271.

[175] Ya.I. Kanel', *Some problems involving burning-theory equations*, Soviet Math. Dokl. **2** (1961), 48–51. Translation of: Dokl. Akad. Nauk SSSR **136** (1961), 277–280.

[176] Ya.I. Kanel', *Stabilization of solutions of the Cauchy problem for equations encountered in combustion theory* (in Russian), Mat. Sb. **59** (1962), 245–288.

[177] T. Kawahara & M. Tanaka, *Interactions of travelling fronts: An exact solution of a nonlinear diffusion equation*, Phys. Lett. A **97** (1983), 311–314.

[178] H. Kawarada, *On solutions of initial-boundary problem for* $u_t = u_{xx} + \frac{1}{1-u}$, Publ. Res. Inst. Math. Sci. **10** (1975), 729–736.

[179] W. Kelley, *Travelling wave solutions of reaction-diffusion equations with convection*, Recent Trends in Differential Equations (edited by R.P. Agarwal), World Scientific Publishing, Singapore, 1992, pp. 355–363.

[180] W. Kelley, *Travelling wave solutions of reaction-diffusion models with density-dependent diffusion*, Rocky Mountain J. Math. **24** (1994), 129–134.

[181] R. Kersner, *The behavior of temperature fronts in media with nonlinear thermal conductivity under absorption*, Moscow Univ. Math. Bull. **33** (1978), 35–41. Translation of: Vestnik Moskov. Univ. Ser. I Mat. Mekh. **33** (1978), 44–51.

[182] R. Kersner, *Filtration with absorption: Necessary and sufficient condition for the propagation of perturbations to have finite velocity*, J. Math. Anal. Appl. **90** (1982), 463–479.

[183] D.A. Kessler, Z. Nir & L.M. Sander, *Front propagation: Precursors, cutoffs, and structural stability*, Phys. Rev. E **58** (1998), 107–114.

[184] N.V. Khusnytdinova, *The limiting moisture profile during infiltration in to a homogeneous soil*, J. Appl. Math. Mech. **31** (1967), 783–789. Translation of: Prikl. Mat. Mekh. **31** (1967), 770–776.

[185] B.F. Knerr, *The porous medium equation in one dimension*, Trans. Amer. Math. Soc. **234** (1977), 381–415.

[186] A. Kolmogoroff, I. Petrovsky & N. Piscounoff, *Étude de l'équation de la diffusion avec croissance de la quantité de matière et son application a un problème biologique*, Bulletin de l'Université d'état à Moscou Série Internationale Section A Mathématiques et Mécanique **1** (1937), 1–25. Appearing in English translation in: Dynamics of Curved Fronts (edited by P. Pelcé), Academic Press, Boston, 1988, pp. 105–130; Selected Works of A.N. Kolmogorov Volume I Mathematics and Mechanics (edited by V.M. Tikhomirov), Kluwer Academic Publishers, Dordrecht, The Netherlands, 1991, pp. 242–270; and; I.G. Petrowsky Selected Works Part II Differential Equations and Probability Theory (edited by O.A. Oleinik), Gordon and Breach Publishers, Amsterdam, 1996, pp. 106–132.

[187] H.-O. Kreiss & J. Lorenz, *Initial-Boundary Value Problems and the Navier-Stokes Equations*, Academic Press, Boston, 1989.

[188] S.N. Kruzhkov, *Generalized solutions of the Cauchy problem in the large for nonlinear equations of first order*, Soviet Math. Dokl. **10** (1969), 785–788. Translation of: Dokl. Akad. Nauk SSSR **187** (1969), 29–32.

[189] S.N. Kruzhkov & N.S. Petrosyan, *Asymptotic behaviour of the solutions of the Cauchy problem for non-linear first order equations*, Russian Math. Surveys **42** (1987), 1–47. Translation of: Uspekhi Mat. Nauk **42** (1987), 3–40.

[190] D.A. Larson, *Transient bounds and time-asymptotic behavior of solutions to nonlinear equations of Fisher type*, SIAM J. Appl. Math. **34** (1978), 93–103.

[191] K.-S. Lau, *On the nonlinear diffusion equation of Kolmogorov, Petrovsky and Piscounov*, J. Differential Equations **59** (1985), 44–70.

[192] J.A. Leach & D.J. Needham, *The evolution of travelling waves in generalized Fisher equations via matched asymptotic expansions: Algebraic corrections*, Quart. J. Mech. Appl. Math. **54** (2001), 157–175.

[193] L.S. Leibenzon, *The motion of Natural Liquids and Gases in Porous Media* (in Russian), OGIS Gosudarstvenoe Izdatel'stva Tekhniko-Teoreticheskoi Literatury, Moscow, 1947.

[194] J.D. Logan, *Weighted-L^2 stability of traveling waves in a porous medium*, Comm. Appl. Nonlinear Anal. **4** (1997), 55–62.

[195] J.D. Logan, *Wave front solutions to a filtration equation with growth*, Comm. Appl. Nonlinear Anal. **5** (1998), 33–43.

[196] B.Q. Lu, B.Z. Xiu, Z.L. Pang & X.F. Jiang, *Exact traveling wave solution of one class of nonlinear diffusion equations*, Phys. Lett. A **175** (1993), 113–115.

[197] L. Malaguti & C. Marcelli, *Travelling wavefronts in reaction-diffusion equations with convection effects and non-regular terms*, Math. Nachr. **242** (2002), 148–164.

[198] W. Malfliet, *Solitary wave solutions of nonlinear wave equations*, Amer. J. Phys. **60** (1992), 650–654.

[199] L.K. Martinson, *The finite velocity of propagation of thermal perturbations in media with constant thermal conductivity*, U.S.S.R. Comput. Math. and Math. Phys. **16** (1976), 141–149. Translation of: Zh. Vychisl. Mat. i Mat. Fiz. **16** (1976), 1233–1241.

[200] C. Mascia, *Travelling wave solutions for a balance law*, Proc. Roy. Soc. Edinburgh Sect. A **127** (1997), 567–593.

[201] H.P. McKean, *Nagumo's equation*, Adv. Math. **4** (1970), 209–223.

[202] H.P. McKean, *Application of Brownian motion to the equation of Kolmogorov-Petrovskii-Piskunov*, Comm. Pure Appl. Math. **28** (1975), 323–331.

[203] G.S. Medvedev, K. Ono & P.J. Holmes, *Travelling wave solutions of the degenerate Kolmogorov-Petrovski-Piskunov equation*, European J. Appl. Math. **14** (2003), 343–367.

[204] R.K. Miller, *Nonlinear Volterra Integral Equations*, W.A. Benjamin, Menlo Park, California, 1971.

[205] A.R. Mitchell, *Numerical studies of travelling waves in nonlinear diffusion equations*, Bull. IMA **17** (1981), 14–20.

[206] A.R. Mitchell & L.B. Reid, *A Galerkin method with quadrature for non linear parabolic equations in population genetics*, The Mathematics of Finite Elements and Applications VI (edited by J.R. Whiteman), Academic Press, London, 1988, pp. 179–190.

[207] H.J.K. Moet, *A note on the asymptotic behavior of solutions of the KPP equation*, SIAM J. Math. Anal. **10** (1979), 728–732.

[208] D. Mollison, *Spatial contact models for ecological and epidemic spread*, J. Roy. Statist. Soc. Ser. B **39** (1977), 283–326.

[209] P. de Mottoni, *Aggregation effects in degenerate diffusion problems*, Japan J. Appl. Math. **1** (1984), 67–91.

[210] J.D. Murray, *Lectures on Nonlinear-Differential-Equation Models in Biology*, Clarendon Press, Oxford, 1977.

[211] J.D. Murray, *Mathematical Biology*, Springer-Verlag, Berlin, 1989.

[212] M. Muskat, *The Flow of Homogeneous Fluids through Porous Media*, McGraw-Hill Book Company, New York, 1937.

[213] J. Nagumo, S. Arimoto & S. Yoshizawa, *An active pulse transmission line simulating nerve axon*, Proc. Inst. Radio Engrs. **50** (1960), 2061–2070.

[214] J. Nagumo, S. Yoshizawa & S. Arimoto, *Bistable transmission lines*, IEEE Trans. Circuit Theory **12** (1965), 400–412.

[215] D.J. Needham & A.N. Barnes, *Reaction-diffusion and phase waves occurring in a class of scalar reaction-diffusion equations*, Nonlinearity **12** (1999), 41–58.

[216] A.C. Newell & J.A. Whitehead, *Finite bandwidth, finite amplitude convection*, J. Fluid Mech. **38** (1969), 279–303.

[217] A.C. Newell & J.A. Whitehead, *Review of the finite bandwidth concept*, Instability of Continuous Systems (edited by H. Leipholz), Springer-Verlag, Berlin, 1971, pp. 284–289.

[218] W.I. Newman, *Some exact solutions to a non-linear diffusion problem in population genetics and combustion*, J. Theoret. Biol. **85** (1980), 325–334.

[219] W.I. Newman, *The long-time behavior of the solution to a non-linear diffusion problem in population genetics and combustion*, J. Theoret. Biol. **104** (1983), 473–484.

[220] W.I. Newman & C. Sagan, *Galactic civilizations: Population dynamics and interstellar diffusion*, Icarus **46** (1981), 293–327.

[221] G. Nicolis & I. Prigogine, *Self-Organization in Nonequilibrium Systems*, John Wiley & Sons, New York, 1977.

[222] A. Okubo, *Diffusion and Ecological Problems: Mathematical Models*, Springer-Verlag, Berlin, 1980.

[223] O.A. Oleinik, A.S. Kalashnikov & Chzhou Y.-L., *The Cauchy problem and boundary problems for equations of the type of non-stationary filtration* (in Russian), Izv. Akad. Nauk SSSR Ser. Mat. **22** (1958), 667–704.

[224] S. Osher & J. Ralston, *L^1 stability of travelling waves with applications to convective porous media flow*, Comm. Pure Appl. Math. **35** (1982), 737–749.

[225] M. Otwinkowski, R. Paul & W.G. Laidlaw, *Exact travelling wave solutions of a class of nonlinear diffusion equations by reduction to a quadrature*, Phys. Lett. A **128** (1988), 483–487.

[226] A. de Pablo & A. Sánchez, *Travelling wave behaviour for a porous-Fisher equation*, European J. Appl. Math. **9** (1998), 285–304.

[227] A. de Pablo & A. Sánchez, *Global travelling waves in reaction-convection-diffusion equations*, J. Differential Equations **165** (2000), 377–413.

[228] A. de Pablo & J.L. Vazquez, *Travelling waves and finite propagation in a reaction-diffusion equation*, J. Differential Equations **93** (1991), 19–61.

[229] A. de Pablo & J.L. Vazquez, *An overdetermined initial and boundary-value problem for a reaction-diffusion equation*, Nonlinear Anal. **19** (1992), 259–269.

[230] C.V. Pao, *Nonlinear Parabolic and Elliptic Equations*, Plenum Press, New York, 1992.

[231] K. Parliński & P. Zieliński, *On the kinetics of the first order phase transitions*, Z. Phys. B **44** (1981), 317–323.

[232] J.P. Pauwelussen & L.A. Peletier, *Clines in the presence of asymmetric migration*, J. Math. Biology **11** (1981), 207–233.

[233] S.L. Paveri-Fontana, *Traveling waves for the enclosed atomizer*, Math. Model. **8** (1987), 275–278.

[234] L.A. Peletier, *Asymptotic stability of travelling waves*, Instability of Continuous Systems (edited by H. Leipholz), Springer-Verlag, Berlin, 1971, pp. 418–422.

[235] L.A. Peletier, *The porous media equation*, Applications of Nonlinear Analysis in the Physical Sciences (edited by H. Amann, N. Bazley & K. Kirchgässner), Pitman Advanced Publishing Program, Boston, 1981, pp. 229–241.

[236] J.R. Philip, *A very general class of exact solutions in concentration-dependent diffusion*, Nature **185** (1960), 233.

[237] J.R. Philip, *General method of exact solution of the concentration-dependent diffusion equation*, Austral. J. Phys. **13** (1960), 1–12.

[238] J.R. Philip, *Theory of infiltration*, Advances in Hydroscience Volume 5 (edited by Ven Te Chow), Academic Press, New York, 1969, pp. 215–296.

[239] D. Phillips, *Existence of solutions of quenching problems*, Applicable Anal. **24** (1987), 253–264.

[240] J. Powell & M. Tabor, *Non-generic connections corresponding to front solutions*, J. Phys. A **25** (1992), 3773–3796.

[241] J.A. Powell, *Conditional stability of front solutions*, J. Math. Biol. **35** (1997), 729–747.

[242] J.A. Powell, A.C. Newell & C.K.R.T. Jones, *Competition between generic and nongeneric fronts in envelope equations*, Phys. Rev. A **44** (1991), 3636–3652.

[243] H. Rhee, B.F. Bodin & N.R. Amundson, *A study of the shock layer in equilibrium exchange systems*, Chem. Engrg. Sci **26** (1971), 1571–1580.

[244] H.-K. Rhee & N.R. Amundson, *A study of the shock layer in nonequilibrium exchange systems*, Chem. Engrg. Sci **27** (1972), 199–211.

[245] H.-K. Rhee & N.R. Amundson, *Shock layer in two solute chromatography: Effect of axial dispersion and mass transfer*, Chem. Engrg. Sci **29** (1974), 2049–2060.

[246] L.A. Richards, *Capillary conduction of liquids through porous mediums*, Physics **1** (1931), 318–333.

[247] P. Rosenau, *Reaction and concentration dependent diffusion model*, Phys. Rev. Lett. **88** (2002) 194501-1–194501-4.

[248] S. Rosseland, E. Jensen & E. Tandberg-Hanssen, *Some considerations on thermal conduction and magnetic fields in prominences*, Electromagnetic Phenomena in Cosmical Physics (edited by B. Lehnert), Cambridge University Press, Cambridge, 1958, pp. 150–157.

[249] F. Rothe, *Convergence to travelling fronts in semilinear parabolic equations*, Proc. Roy. Soc. Edinburgh Sect. A **89** (1978), 213–234.

[250] W. van Saarloos, *Dynamical velocity selection: Marginal stability*, Phys. Rev. Lett. **58** (1987), 2571–2574.

[251] W. van Saarloos, *Front propagation into unstable states: Marginal stability as a dynamical mechanism for velocity selection*, Phys. Rev. A **37** (1988), 211–229.

[252] W. van Saarloos, *Front propagation into unstable states. II. Linear versus nonlinear marginal stability and rate of convergence*, Phys. Rev. A **39** (1989), 6367–6390.

[253] W. van Saarloos, M. van Hecke & R. Hołyst, *Front propagation into unstable and metastable states in smetic-C* liquid crystals: Linear and nonlinear marginal-stability analysis*, Phys. Rev. E **52** (1995), 1773–1777.

[254] F. Sánchez-Garduño & P.K. Maini, *Existence and uniqueness of a sharp travelling wave in degenerate non-linear diffusion Fisher-KPP equations*, J. Math. Biol. **33** (1994), 163–192.

[255] F. Sánchez-Garduño & P.K. Maini, *An approximation to a sharp type solution of a density-dependent reaction-diffusion equation*, Appl. Math. Lett. **7** (1994), 47–51.

[256] F. Sánchez-Garduño & P.K. Maini, *Travelling wave phenomena in some degenerate reaction-diffusion equations*, J. Differential Equations **117** (1995), 281–319.

[257] F. Sánchez-Garduño & P.K. Maini, *Travelling wave phenomena in non-linear diffusion degenerate Nagumo equations*, J. Math. Biol. **35** (1997), 713–728.

[258] F. Sánchez-Garduño, P.K. Maini & E. Kappos, *A review of travelling wave solutions of one-dimensional reaction-diffusion equations with non-linear diffusion term*, Forma **11** (1996), 45–59.

[259] F. Sánchez-Garduño, P.K. Maini & M.E. Kappos, *A shooting argument approach to a sharp-type solution for nonlinear degenerate Fisher-KPP equations*, IMA J. Appl. Math. **57** (1996), 211–221.

[260] J. Satsuma, *Explicit solutions of nonlinear equations with density dependent diffusion*, J. Phys. Soc. Japan **56** (1987), 1947–1950.

[261] J. Satsuma, *Exact solutions of Burgers' equation with reaction terms*, Topics in Soliton Theory and Exactly Solvable Nonlinear Equations (edited by M. Ablowitz, B. Fuchssteiner & M. Kruskal), World Scientific Publishing, Singapore, 1987, pp. 255–262.

[262] D.H. Sattinger, *On the stability of waves of nonlinear parabolic systems*, Adv. Math. **22** (1976), 312–355.

[263] D.H. Sattinger, *Weighted norms for the stability of traveling waves*, J. Differential Equations **25** (1977), 130–144.

[264] A. Schatz, *Free boundary problems of Stephan type with prescribed flux*, J. Math. Anal. Appl. **28** (1969), 569–580.

[265] H. Schlichting, *Boundary-Layer Theory*, McGraw-Hill Book Company, New York, 1968.

[266] A.C. Scott, *The electrophysics of a nerve fiber*, Rev. Modern Phys. **47** (1975), 487–533.

[267] A.C. Scott, *Neurophysics*, John Wiley & Sons, New York, 1977.

[268] S.K. Scott & K. Showalter, *Simple and complex propagating reaction-diffusion fronts*, J. Phys. Chem. **96** (1992), 8702–8711.

[269] L.A. Segel, *Taxes in cellular ecology*, Mathematical Ecology (edited by S.A. Levin & T.G. Hallam), Lecture Notes in Biomathematics **54**, Springer-Verlag, Berlin, 1984, pp. 407–424.

[270] K.G. Shkadinskii, V.V. Barelko & I.I. Kurochka, *Set of solutions in the form of traveling waves for the combustion-type equation*, Dokl. Phys. Chem. **233** (1977), 377–379. Translation of: Dokl. Akad. Nauk SSSR **233** (1977), 639–641.

[271] W. Simon, P. Reichert & C. Hinz, *Properties of exact and approximate traveling wave solutions for transport with nonlinear and nonequilibrium sorption*, Water Resources Res. **33** (1997), 1139–1147.

[272] B.D. Sleeman & E. Tuma, *On exact solutions of a class of reaction-diffusion equations*, IMA J. Appl. Math. **33** (1984), 153–168.

[273] J. Smoller, *Shock Waves and Reaction-Diffusion Equations*, Springer-Verlag, New York, 1983.

[274] Song B., *Existence, uniqueness and properties of the solutions of a degenerate parabolic equation with diffusion-advection-absorption*, Acta Math. Appl. Sinica (English Ser.) **10** (1994), 113–140.

[275] I.W. Stewart, T. Carlsson & F.M. Leslie, *Chaotic instabilities in smetic-C liquid crystals*, Phys. Rev. E **49** (1994), 2130–2140.

[276] A.N. Stokes, *On two types of moving front in quasilinear diffusion*, Math. Biosci. **31** (1976), 307–315.

[277] A.N. Stokes, *Nonlinear diffusion waveshapes generated by possibly finite initial disturbances*, J. Math. Anal. Appl. **61** (1977), 370–381.

[278] D. Swartzendruber, *The flow of water in unsaturated soils*, Flow through Porous Media (edited by R.J.M. de Wiest), Academic Press, New York, 1969, pp. 215–292.

[279] J.E. Taylor, J.W. Cahn & C.A. Handwerker, *Geometric models of crystal growth*, Acta Metallurgica Materialia **40** (1992), 1443–1474.

[280] J.K. Tegnér, *Properties of simple model problems for reacting gas flows*, Math. Methods Appl. Sci. **17** (1994), 577–594.

[281] D. Terman, *Traveling wave solutions of multistable reaction-diffusion equations*, Contemp. Math. **17** (1983), 361–378.

[282] D. Terman, *Directed graphs and traveling waves*, Trans. Amer. Math. Soc. **289** (1985), 809–847.

[283] K. Uchiyama, *The behavior of solutions of the equation of Kolmogorov-Petrovsky-Piskunov*, Proc. Japan Acad. Ser. A Math. Sci. **53** (1977), 225–228.

[284] K. Uchiyama, *The behavior of solutions of some non-linear diffusion equations for large time*, J. Math. Kyoto Univ. **18** (1978), 453–508.

[285] K. Uchiyama, *Brownian first exit from and sojourn over one sided moving boundary and application*, Z. Wahrsch. Verw. Gebiete **54** (1980), 75–116.

[286] K. Uchiyama, *Asymptotic behavior of solutions of reaction-diffusion equations with varying drift coefficients*, Arch. Rational Mech. Anal. **90** (1985), 291–311.

[287] M. Ughi, *On the porous media equation with either source or absorption*, Cuadernos del Instituto de Matemática "Beppo Levi" **22** (1991), 1–72.

[288] V. Vanaja & P.L. Sachdev, *Asymptotic solutions of a generalized Burgers equation*, Quart. Appl. Math. **50** (1992), 627–640.

[289] J.L. Vazquez, *An introduction to the mathematical theory of the porous medium equation*, Shape Optimization and Free Boundaries (edited by M.C. Delfour & G. Sabidussi), Kluwer Academic Publishers, Dordrecht, The Netherlands, 1992, pp. 347–389.

[290] E.J.M. Veling, *Travelling waves in an initial-boundary value problem*, Proc. Roy. Soc. Edingburgh Sect. A **90** (1981), 41–61.

[291] E.J.M. Veling, *Pushed travelling waves in an initial-boundary value problem for Fisher type equations*, Nonlinear Anal. **6** (1982), 1271–1286.

[292] G. Verbist & D. Weaire, *A soluble model for foam drainage*, Europhys. Lett. **26** (1994), 631–634.

[293] G. Verbist, D. Weaire & A.M. Kraynik, *The foam drainage equation*, J. Phys. Condensed Matter **8** (1996), 3715–3731.

[294] P.-F. Verhulst, *Notice sur la loi que la population suit dans son accroissement*, Correspondance Mathématique et Physique **10** (1838), 113–121.

[295] A.I. Vol'pert, *Application of functionals to the investigation of travelling waves* (in Russian), Operator Equations and Set Functions (edited by S.I. Khudyaev), Perm. Gos. Univ., Syktyvkar, 1985, pp. 30–35.

[296] A.I. Volpert, *Propagation of waves described by nonlinear parabolic equations*, I.G. Petrowsky Selected Works Part II Differential Equations and Probability Theory (edited by O.A. Oleinik), Gordon and Breach Publishers, Amsterdam, 1996, pp. 364–399.

[297] A.I. Volpert, V.A Volpert & V.A. Volpert, *Traveling Wave Solutions of Parabolic Systems*, American Mathematical Society, Providence, R.I., 1994.

[298] V.A. Vol'pert, *Asymptotic behavior of solution of a nonlinear diffusion equation with a source term of general form*, Siberian Math. J. **30** (1989), 25–36. Translation of: Sibirsk. Mat. Zh. **30** (1989), 35–47.

[299] Wang J., *Weak travelling wave front solutions of generalized diffusion equations with reaction*, Chin. Ann. Math. Ser. B **15** (1994), 283–292.

[300] Wang M., *Travelling wave solutions of degenerate parabolic equations* (in Chinese), Chinese Ann. Math. Ser. A **12** (1991), 627–635.

[301] M. Wang, S. Xiong & Q. Ye, *Explicit wave front solutions of Noyes-Field systems for the Belousov-Zhabotinskii reaction*, J. Math. Anal. Appl. **182** (1994), 705–717.

[302] X.Y. Wang, *Exact and explicit solitary wave solutions for the generalised Fisher equation*, Phys. Lett. A **131** (1988), 277–279.

[303] X.Y. Wang, Z.S. Zhu & Y.K. Lu, *Solitary waves solutions of the generalised Burgers-Huxley equation*, J. Phys. A **23** (1990), 271–274.

[304] D. Weaire, S. Hutzler, G. Verbist & E. Peters, *A review of foam drainage*, Advances in Chemical Physics Volume 102 (edited by I. Prigogine & S.A. Rice), John Wiley & Sons, New York, 1997, pp. 315–374.

[305] T.P. Witelski, *An asymptotic solution for traveling waves of a nonlinear-diffusion Fisher's equation*, J. Math. Biol. **33** (1994), 1–16.

[306] J. Xin, *Front propagation in heterogeneous media*, SIAM Rev. **42** (2000), 161–230.

[307] J.X. Xin, *Existence of multidimensional traveling waves in the transport of reactive solutes through periodic porous media*, Arch. Rational Mech. Anal. **128** (1994), 75–103.

[308] J.X. Xin, *Stabilty of traveling waves of a solute transport equation*, J. Differential Equations **135** (1997), 269–298.

[309] X. Xin, *Existence and uniqueness of travelling waves in a reaction-diffusion equation with combustion nonlinearity*, Indiana Univ. Math. J. **40** (1991), 985–1008.

[310] Z.J. Yang, *Travelling wave solutions to nonlinear evolution and wave equations*, J. Phys. A **27** (1994), 2837–2855.

[311] Z.J. Yang, R.A. Dunlap & D.J.W. Geldart, *Exact traveling wave solutions to nonlinear diffusion and wave eqautions*, Internat. J. Theoret. Phys. **33** (1994), 2057–2065.

[312] S.E.A.T.M. van der Zee, *Analytical traveling wave solutions for transport with nonlinear and nonequilibrium adsorption*, Water Resources Res. **26** (1990), 2563–2578, ibid. **27** (1991), 983.

[313] S.E.A.T.M. van der Zee & W.H. van Riemsdijk, *Transport of reactive solute in spatially variable soil systems*, Water Resources Res. **23** (1987), 2059–2069.

[314] Y.B. Zeldovich, *Theory of Flame Propagation*, National Advisory Committee for Aeronautics Technical Memorandum 1282 (1951), 39 pp. Translation of: Zh. Fiz. Khim. **22** (1948), 27–49.

[315] Ya.B. Zeldovich, G.I. Barenblatt, V.B. Librovich & G.M. Makhviladze, *The Mathematical Theory of Combustion and Explosions*, Consultants Bureau, New York, 1985.

[316] Ya.B. Zel'dovich & Yu.P. Raizer, *Physics of Shock Waves and High-Temperature Hydrodynamic Phenomena Volume I*, Academic Press, New York, 1966.

[317] Ya.B. Zel'dovich & Yu.P. Raizer, *Physics of Shock Waves and High-Temperature Hydrodynamic Phenomena Volume II*, Academic Press, New York, 1967.

[318] J.B. Zeldowitsch & D.A. Frank-Kamenetzki, *A theory of thermal propagation of flame*, Dynamics of Curved Fronts (edited by P. Pelcé), Academic Press, Boston, 1988, pp. 131–140. Translation from: Acta Physicochimica U.R.S.S. **9** (1938).

Index

For orders originating from all over the world except USA/Canada/Latin America:

Birkhäuser Verlag AG
c/o Springer GmbH & Co
Haberstrasse 7
D-69126 Heidelberg
Fax: +49 / 6221 / 345 4 229
e-mail: birkhauser@springer.de
http://www.birkhauser.ch

MIX
Papier aus verantwortungsvollen Quellen
Paper from responsible sources
FSC® C105338

If you have any concerns about our products,
you can contact us on
ProductSafety@springernature.com

In case Publisher is established outside the EU,
the EU authorized representative is:
Springer Nature Customer Service Center GmbH
Europaplatz 3, 69115 Heidelberg, Germany

Printed by Libri Plureos GmbH
in Hamburg, Germany